The physics of atmospheres

The physics
of atmospheres

JOHN T. HOUGHTON CBE, FRS

Director General, Meteorological Office

SECOND EDITION

CAMBRIDGE
UNIVERSITY PRESS

Published by the Press Syndicate of the University of Cambridge
The Pitt Building, Trumpington Street, Cambridge CB2 1RP
40 West 20th Street, New York, NY 10011–4211, USA
10 Stamford Road, Oakleigh, Melbourne 3166, Australia

First published 1977
First paperback edition 1979
Second edition 1986
Reprinted 1989, 1991, 1995

Printed in Great Britain by J W Arrowsmith Ltd, Bristol

British Library cataloguing in publication data
Houghton, John T.
   The physics of atmospheres. – 2nd ed.
   1. Atmosphere
   I. Title
   551.5   QC880

Library of Congress cataloguing in publication data
Houghton, John Theodore.
   The physics of atmospheres.

   Bibliography
   Includes index.
   1. Atmospheric physics.   2. Dynamic meteorology.
I. Title.
QC880.H68   1986   551.5′01′53   86-6875

ISBN 0 521 33956 1 paperback
(First edition ISBN 0 521 21443 2 hard covers
               ISBN 0 521 29656 0 paperback)

MP

To Janet and Peter

The works of the Lord are great, sought out of all them
that have pleasure therein.
Psalm 111, v. 2.

Contents

Preface to first edition   xiii
Preface to second edition   xv
Acknowledgements   xvi

1   Some basic ideas   1
1.1   Planetary atmospheres   1
1.2   Equilibrium temperatures   2
1.3   Hydrostatic equation   3
1.4   Adiabatic lapse rate   4
1.5   Sandström's theorem   5
      Problems   7

2   A radiative equilibrium model   9
2.1   Black-body radiation   9
2.2   Absorption and emission   10
2.3   Radiative equilibrium in a grey atmosphere   12
2.4   Radiative time constants   15
2.5   The greenhouse effect   15
      Problems   17

3   Thermodynamics   19
3.1   Entropy of dry air   19
3.2   Vertical motion of saturated air   19
3.3   The tephigram   22
3.4   Total potential energy of an air column   23
3.5   Available potential energy   24

3.6	Zonal and eddy energy	29
	Problems	30
4	**More complex radiation transfer**	39
4.1	Solar radiation: its modification by scattering	39
4.2	Absorption of solar radiation by ozone	40
4.3	Absorption by single lines	41
4.4	Transmission of an atmospheric path	44
4.5	The integral equation of transfer	45
4.6	Integration over frequency	47
4.7	Heating rate due to radiative processes	48
4.8	Cooling by carbon dioxide emission from upper stratosphere and lower mesosphere	48
4.9	Band models	49
4.10	Continuum absorption	50
4.11	Global radiation budget	50
	Problems	52
5	**The middle and upper atmospheres**	57
5.1	Temperature structure	57
5.2	Diffusive separation	58
5.3	The escape of hydrogen	59
5.4	The energy balance of the thermosphere	65
5.5	Photochemical processes	66
5.6	Breakdown of thermodynamic equilibrium	70
	Problems	76
6	**Clouds**	80
6.1	Cloud formation	80
6.2	The growth of cloud particles	80
6.3	The radiative properties of clouds	82
6.4	Radiative transfer in clouds	83
	Problems	85
7	**Dynamics**	88
7.1	Total and partial derivatives	88
7.2	Equations of motion	89
7.3	The geostrophic approximation	92
7.4	Cyclostrophic motion	93

7.5	Surfaces of constant pressure	93
7.6	The thermal wind equation	94
7.7	The equation of continuity	95
	Problems	97
8	**Atmospheric waves**	105
8.1	Introduction	105
8.2	Sound waves	105
8.3	Gravity waves	106
8.4	Rossby waves	111
8.5	The vorticity equation	113
8.6	Three dimensional Rossby-type waves	115
	Problems	118
9	**Turbulence**	126
9.1	The Reynolds number	126
9.2	Reynolds stresses	127
9.3	Ekman's solution	129
9.4	The mixing-length hypothesis	131
9.5	Ekman pumping	132
9.6	The spectrum of atmospheric turbulence	133
	Problems	136
10	**The general circulation**	138
10.1	Laboratory experiments	138
10.2	A symmetric circulation	140
10.3	Inertial instability	143
10.4	Barotropic instability	145
10.5	Baroclinic instability	147
10.6	Sloping convection	150
10.7	Energy transport	151
10.8	Transport of angular momentum	153
10.9	The general circulation of the middle atmosphere	156
	Problems	159
11	**Numerical modelling**	165
11.1	A barotropic model	165
11.2	Baroclinic models	166
11.3	Primitive equation models	168

11.4   Inclusion of orography   169
11.5   Convection   169
11.6   Moist processes   170
11.7   Radiation transfer   171
11.8   Inclusion of clouds   174
11.9   Sub grid scale processes   175
11.10   Transfer across the surface   176
11.11   Forecasting models   177
11.12   Other models   179
        Problems   181

12   Global observation   189
12.1   What observations are required?   189
12.2   Conventional observations   190
12.3   Remote sounding from satellites   192
12.4   Remote sounding of atmospheric temperature   193
12.5   Remote measurements of composition   199
12.6   Other remote sounding observations   201
12.7   Observations from remote platforms   203
12.8   Achieving global coverage   203
        Problems   205

13   Atmospheric predictability and climatic change   210
13.1   Short-term predictability   210
13.2   Variations of climate   212
13.3   Atmospheric feedback processes   213
13.4   Different kinds of predictability   215
13.5   Jupiter's Great Red Spot   216
13.6   The challenge of climate research   218
        Problems   219

        Appendices   224
1   Some useful physical constants and data on dry air   224
2   Properties of water vapour   225
3   Atmospheric composition   226
4   Relation of geopotential to geometric height   227
5   Model atmospheres (0–105 km)   227
6   Mean reference atmosphere (110–500 km)   236
7   The Planck function   236

Contents

8	Solar radiation	238
9	Absorption of solar radiation by oxygen and ozone	240
10	Spectral band information	242
	Bibliography	252
	References to works cited in the text	255
	Answers to problems and hints to their solution	259
	Index	265

Preface

During the last ten years or so, three important factors have combined to bring about a large increase of interest in atmospheric science. First, man's increasing industrial activity has brought into much sharper focus the problems of pollution and the possibility of artificial modification of the environment either inadvertently or in a controlled way. Secondly, concern about world food resources in the face of rapidly increasing population has made us much more aware of the critical effect of fluctuations in climate, particularly in parts of the developing world. Thirdly, the advent of the artificial satellite and developments in electronic computers have made available much more powerful tools for atmospheric research and have made it possible to study the atmosphere as a whole.

At the present time, study of the global atmosphere is being concentrated by a large international programme – the Global Atmospheric Research Programme – which is being organized jointly by atmospheric scientists (through the International Council of Scientific Unions) and the national weather services (through the World Meteorological Organization). The aims of this programme are to attack some of the basic problems regarding our understanding of the behaviour and circulation of the whole atmosphere (hence the term global) and of the mechanisms of climatic change.

It is in the context of this global programme that this book has been written. Its aim is to introduce physics students at both undergraduate and graduate levels to the physical processes which govern the structure and circulation of a planetary atmosphere. In writing the book I have concentrated on the basic physics and wherever possible have constructed simple physical models by applying to the atmosphere the principles of classical thermodynamics, radiation transfer and fluid mechanics.

The book is meant to be a working text and for that reason includes a large number of problems. Some of the problems extend considerably the basic material in the chapters; other problems are included to illustrate what is already in the text.

I have attempted to introduce all the main areas of study involved in the physics of the neutral atmosphere (the ionized atmosphere and the magnetosphere have been deliberately omitted) while at the same time keeping the book within reasonably small compass. This has necessitated a rigorous selection of material. I hope, however, I have achieved a balanced text from which the reader will be able to acquire a perspective of the atmosphere as a whole and a feeling for where the subject is going. A substantial bibliography is included to help the student to pursue further reading. Also included are a number of appendices summarizing information concerning the atmosphere which research workers will find useful for reference.

Because atmospheric science cuts across a number of traditional disciplines in physics and chemistry, symbols nomenclature and units can be even more of a problem than usual. On the question of symbols I have tended to use those familiar in the various subjects rather than use unfamiliar ones in the interests of uniformity. I have, in general, employed SI units, with one notable exception. For pressure I have kept to the use of millibars, which are universally used in operational meteorology, rather than change to Pascals.

I am indebted to many colleagues and students from whom I have received help in writing the book. Some of the ideas and material for the dynamical chapters arose from an admirable series of lectures on Atmospheric Dynamics given in Oxford by Dr R. S. Harwood, who also made valuable comments and criticisms regarding the text of those chapters. Other criticisms and suggestions on various parts of the text were made by Dr C. D. Walshaw, Dr G. D. Peskett, Professor R. Hide, Professor P. A. Sheppard and by graduate students of Oxford University who have used the text and done the problems. Dr C. D. Rodgers, Mrs M. Corney and Mr R. J. Wells assisted in the collection of some of the material for the Appendices, Mr M. J. Wale has assisted with providing the answers to problems and hints to their solutions. Mr D. Wrigley has read through the proofs and provided the index. I also wish particularly to thank Miss C. M. Wagstaff who typed the manuscript, Miss B. Linder who drew most of the diagrams, and the staff of the Cambridge University Press for their courtesy and assistance during the preparation of the book.

March 1976   J. T. Houghton

Preface to the second edition

The opportunity of a new edition has been taken to make significant additions and alterations to some of the chapters. In particular, chapter 4 on radiative transfer has been restructured to give a more logical development and chapters 10, 11, 12 and 13 on the general circulation, numerical modelling, observations and climate have been significantly extended to take account of modern developments and increasing interest in these topics.

The inclusion of a large number of problems as a didactic aid has been much appreciated. In this new edition nearly thirty new problems have been added, many of them containing a substantial amount of exposition in order to illustrate principles or to introduce new ideas.

In order to come into line with recommended practice, the unit of pressure throughout this new edition is the Pascal.

I am indebted to many colleagues in the Meteorological Office and elsewhere who have suggested improvements for the new edition and assisted in its preparation. I acknowledge particular assistance from Dr D. G. Andrews, Dr K. A. Browning, Dr M. Cullen, Dr R. Hide and Dr F. W. Taylor who have provided suggestions for (and criticism of) the text or the illustrative material and from Mrs Brenda Bell who typed much of the additional text.

January 1986   J. T. Houghton

Acknowledgements

Permission has been received to use published diagrams as a basis for figures in the text from the following: Professors J. Charney, H. U. Dütsch, R. Hide, A. S. Monin and E. Palmén; Drs M. Berenger, K. A. Browning, N. Gerbier, R. Hanel, L. Hembree, P. R. Julian, R. A. McClatchey, S. Manabe, P. J. Mason, L. T. Matveev, C. W. Newton, C. Ridley, D. Rodgers, B. Schlachman, J. E. A. Selby, W. L. Smith, F. W. Taylor, L. W. Uccellini, D. Vanous, W. M. Washington, and R. T. Wetherald; Academic Press Inc.; Akademie-Verlag, Berlin; Air Force Geophysics Laboratory; American Meteorological Society; D. Reidel Publishing Co.; Israel Programme for Scientific Translations; McGraw-Hill Book Co.; Meteorological Office, National Aeronautics and Space Administration, U.S.A.; National Oceanographic and Atmospheric Administration, U.S.A.; Optical Society of America; Royal Meteorological Society; Taylor & Francis Ltd.; University of Dundee Electronics Laboratory and the World Meteorological Organization.

1

Some basic ideas

1.1 Planetary atmospheres

The atmosphere of a planet is the gaseous envelope surrounding it. Large differences exist between the atmospheres of different planets both in chemical composition and physical structure as will be seen from the information about the atmospheres of our nearest planetary neighbours contained in table 1.1.

For a complete understanding of an atmosphere we need to know about its evolution and the processes which have determined its mass and its composition. We also need to know about its physical structure and the distribution of density, composition and motion within the atmosphere.

Table 1.1

	Mean surface temperature (K)	Surface pressure (atm)[a]	Accn due to gravity (m s^{-2})	Main constituents	
Venus	750	90	8.84	>90% CO_2	Deep clouds complete cover
Earth	280	1	9.81	N_2 78%[b] O_2 21%	~50% cover H_2O clouds
Mars	240	0.007	3.76	>80% CO_2	Some very thin H_2O clouds
Jupiter	134[c]	2[c]	26	H_2, He	NH_3 clouds

[a] 1 atm is mean pressure at earth's surface $= 1.013 \times 10^5$ Pa(N m^{-2}) $= 101.3$ kPa $= 1013$ millibars (mb).
[b] See appendix 3 for detailed composition of earth's atmosphere.
[c] At cloud top.

In the case of the earth's atmosphere very detailed information is required in order to predict its state for periods of days or weeks ahead. Further, because of our concern about climatic change, it is necessary to understand the factors which determine the average state of the atmosphere over periods of years and centuries.

In this book discussion will be confined to a description of the physical processes involved in atmospheric study, with application mostly to the earth's atmosphere. Using basic physical principles in thermodynamics, radiation transfer and fluid dynamics, we shall first develop simple models which may be used to give some understanding of the main features of atmospheric structure and behaviour. In later chapters, methods of observing the atmosphere will be described and an indication given of the major problems which are being tackled in atmospheric science at the present time.

1.2   Equilibrium temperatures

A crude estimate of the effective temperature T_e of a planet's surface may be made by equating the solar radiation it absorbs to the infrared radiation it emits as follows:

$$4\pi a^2 \sigma T_e^4 = \pi a^2 (1 - A)F/R^2 \tag{1.1}$$

where σ is the Stefan–Boltzmann constant, a is the radius of the planet at distance R (astronomical units) from the sun, $F (= 1370 \text{ W m}^{-2})$ is the solar flux on a surface near the earth (i.e. when $R = 1$) normal to the solar beam and A is the *aibedo*, i.e. the ratio of reflected to incident solar energy for the whole planet. The right hand side of (1.1) is the absorbed solar radiation and the left hand side the radiation emitted by the planet assuming it behaves as a black body at temperature T_e. Table 1.2 lists values of the various parameters in (1.1) for some of the planets and compares approximate measured temperatures T_m with those calculated from (1.1). The agreement is good except in the case of Jupiter,

Table 1.2

	R	A	T_e (K)	T_m (K)	M_r	Period of rotation (days)
Venus	0.72	0.77	227	230	44	243
Earth	1.00	0.30	256	250	28.8	1.00
Mars	1.52	0.15	216	220	44	1.03
Jupiter	5.20	0.58	98	130	2	0.41

for which planet absorption of solar radiation accounts for only about half the energy input required to maintain its observed temperature; the other half must, therefore, be internally generated (cf. problem 1.2).

The effective temperature for Venus listed in table 1.2 is very different from the temperature at the planet's surface listed in table 1.1. The very dense and complete cloud cover of Venus is substantially opaque to radiation of all wavelengths shorter than about 1 mm. These clouds act as a radiation blanket preventing almost all radiation emitted from the lower atmosphere from escaping, while allowing a small amount of solar radiation through (see § 1.5). This process whereby a high surface temperature can be maintained is sometimes known as the *greenhouse effect*. To a more limited extent than on Venus it is also effective for the earth where the average surface temperature approaches 290 K.

1.3 Hydrostatic equation

Because a planet's atmosphere is in the planet's gravitational field, its density will fall with altitude. Since vertical motion is generally very small, the assumption of static equilibrium is a good starting point. If ρ is the density and p the pressure at altitude z measured vertically upwards from the surface we have

$$dp = -g\rho dz \qquad (1.2)$$

Since a planet's atmosphere is generally of total depth small compared with the planet's radius, the acceleration due to gravity g is approximately constant within the atmospheric region (cf. problem 1.5).

From the equation of state for a perfect gas of molecular weight M_r (see table 1.2) and temperature T,

$$\rho = \frac{M_r p}{RT} \qquad (1.3)$$

where R is the gas constant per mole.

Equation (1.2), therefore, becomes

$$dp/p = -dz/H$$

which on integration gives, for the pressure p at altitude z,

$$p = p_0 \exp\left\{-\int_0^z dz/H\right\} \qquad (1.4)$$

where p_0 is the pressure at $z = 0$, and $H = RT/M_r g$, known as the scale height, is the increase in altitude necessary to reduce the pressure by a factor e. For the lower atmosphere of the earth H varies between 6 km at $T = 210$ K to 8.5 km at

$T = 290$ K. In the very high atmosphere, the molecular weight is no longer constant and the temperature is much higher, a situation which is discussed in more detail in § 5.2.

1.4   Adiabatic lapse rate

The simplest model of a planetary atmosphere we can make is to assume that it is transparent to all radiation, that it contains no liquid particles and that the temperature of its lower boundary is that of the planet's surface whose mean temperature is determined by the simple calculation of § 1.2. Consider the vertical motion of a 'parcel' at pressure p, temperature T and of specific volume V within such an atmosphere. We shall assume the atmosphere to be in hydrostatic equilibrium described by (1.2). Gravitational forces and buoyancy forces are, therefore, balanced and neither need be included explicitly in the expression for the first law of thermodynamics applied to unit mass which is therefore

$$dq = c_v \, dT + p \, dV \tag{1.5}$$

where c_v is the specific heat at constant volume. Providing no heat enters or leaves the parcel, the motion is adiabatic and the quantity of heat dq is zero.

Differentiation of the equation of state (1.3) (remembering that $1/V = \rho$) gives

$$p \, dV + V \, dp = R \, dT / M_r$$
$$= (c_p - c_v) \, dT \tag{1.6}$$

Fig. 1.1. (a) Illustrating the stability of a temperature profile b.

(b) Illustrating adiabatic lapse rate a; *inversion* condition b, i.e. temperature increasing with height, an inversion occurs, for instance, when surface has cooled more rapidly than the atmosphere; *superadiabatic* condition c which cannot persist for very long and never occurs on a large scale.

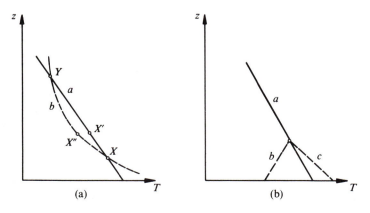

(a)   (b)

since for a perfect gas $c_p - c_v = R/M_r$ where c_p is the specific heat at constant pressure. On substituting for $p\,dV$ from (1.6) in (1.5) we have

$$c_p\,dT - V\,dp = dq = 0 \qquad (1.7)$$

which gives, on substituting from (1.2),

$$\frac{dT}{dz} = -\frac{g}{c_p} = -\Gamma_d \qquad (1.8)$$

Γ_d is known as the *adiabatic lapse rate* for a dry atmosphere. For the earth's atmosphere, $c_p = 1005\ \text{J kg}^{-1}\,\text{K}^{-1}$ and $\Gamma_d \simeq 10\ \text{K km}^{-1}$.

If the atmosphere, therefore, is heated by contact with the surface and vertical motion thereby ensues, we may expect a uniform temperature gradient with altitude of $10\ \text{K km}^{-1}$.

It is informative to consider the stability with respect to vertical motion which occurs in an atmosphere with any temperature distribution b in fig. 1.1. If a parcel initially at X rises adiabatically it will follow the dry adiabatic a to X' where it will be surrounded by atmosphere under the conditions of X'', i.e. it will be warmer than the surroundings and will continue to rise. Point X on curve b is, therefore, unstable. By a similar argument point Y is stable. A given temperature gradient dT/dz represents stable or unstable conditions according as $-dT/dz$ is $<$ or $> \Gamma_d$.

1.5 Sandström's theorem

Instead of assuming that the atmosphere is completely transparent to radiation, let us consider an atmosphere which absorbs strongly, the incident solar energy being deposited in the upper atmosphere, as is likely, for instance, to be the case in the Venus atmosphere where a very deep continuous cloud layer is present. What temperature structure is to be expected in the atmosphere below the absorbing layer? Because increasing temperature with altitude is a very stable situation (§ 1.4) vertical motion will not occur; conduction through the gas and radiation transfer will bring the lower atmosphere to an isothermal state (fig. 1.2).

This situation is an illustration of Sandström's theorem (see A. Defant, 1961) which considers motion in the atmosphere to arise from the operation of a thermodynamic engine with heat sources and sinks. In a simple thermo-dynamic cycle (fig. 1.3), energy is released by going round the cycle in the direction of the arrows. This is only possible if the source is not only at a higher temperature than the sink but also, because for a gas at constant volume pressure increases with temperature, if the source is at a higher pressure than the sink. Hence Sandström's theorem which states that a closed steady circulation can only be maintained in an atmosphere if the heat source is situated at a higher pressure than the heat sink.

Because, in practice, heat is conducted from the precise place where it is deposited to the surrounding atmosphere and the sources and sinks to some extent are distributed by the motions themselves, strict interpretation of the theorem is difficult, as Jeffreys (1925) has pointed out. It is, however, certainly true that strong circulations cannot develop unless the conditions of Sandström's theorem are satisfied.

What then of the lower atmosphere of Venus which is observed to possess an adiabatic lapse rate of temperature from the cloud level to the surface which is at the surprisingly high temperature of ~ 750 K? We can only deduce that sufficient solar energy must penetrate to the surface for enough

Fig. 1. 2. Illustrating an adiabatic lapse rate above and an isothermal state below an absorbing layer.

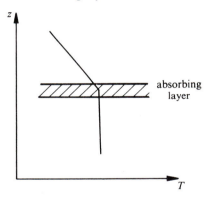

Fig. 1.3. A thermodynamic cycle releases energy when the source at temperature T_1 is at a higher pressure than the sink at temperature T_2.

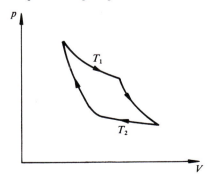

vertical motion to be developed to overcome the effects of conduction and radiation which would tend to make the profile isothermal – a suggestion which is, in fact, in agreement with measurements made by the Russian Venera probes of a small amount of sunlight being still incident on the probes when they reached the Venus surface.

Problems

(See the appendices for lists of constants and useful information about the atmosphere.)

1.1 If the planets of table 1.2 possessed no atmospheres, what would be the equilibrium temperatures of their surfaces, assuming a uniform albedo of the surfaces of 0.05?

1.2 From the information in table 1.2 compute the amount of energy per unit area arising from the internal source on Jupiter.

1.3 If the temperature falls uniformly with height z at 10 K km^{-1}, write down an expression for the fall of pressure with height.

1.4 Calculate for the earth's atmosphere the height of the surface where the pressure is 0.1 of its surface value, assuming (1) uniform temperature of 290 K (2) surface temperature of 290 K and uniform lapse rate of 10 K km^{-1}.

1.5 What is the percentage change of the acceleration due to gravity g between the surface and 100 km altitude? Estimate the error in pressure determination at 100 km latitude through the use of (1.4) if a constant g is assumed.

1.6 The total mass of the oceans is 1.35×10^{21} kg. Compare with this with total mass of the atmosphere. Also, compare the total heat capacity of the atmosphere with that of the oceans.

1.7 From the information in tables 1.1 and 1.2 compute the adiabatic lapse rates for Mars, Venus and Jupiter atmospheres.

1.8 For an unsaturated atmosphere containing 2% by volume of water vapour, find the value of the adiabatic lapse rate and compare it with the value of dry air.

1.9 A balloon filled with helium is required to carry a payload of 100 kg to a height of 30 km. What volume of balloon is required if the material used in its construction is polyethylene sheet, 25×10^{-4} cm thick and of density 1 g cm^{-3}?

1.10 Consider the vertical forces acting on a parcel displaced by a distance Δz from its equilibrium position in an atmosphere where the vertical temperature gradient is dT/dz. Write down the equation of motion in the form

$$\ddot{\Delta z} + gB\,\Delta z = 0$$

where $\quad B = \dfrac{1}{T}\left(\dfrac{dT}{dz} + \Gamma_d\right)$

Discuss solutions to the equation of motion when $B < 0$, $= 0$, > 0. When $B > 0$, show that oscillations occur at the *Brunt–Vaisala frequency* $(gB)^{1/2}$. Calculate its value when $dT/dz = -6.5 \text{ K km}^{-1}$ – an average value for the lower part of the earth's atmosphere. This oscillation is a particular case of a *gravity wave*. These waves will be discussed further in §8.3.

2

A radiative equilibrium model

2.1   Black-body radiation

In fig. 2.1 are two curves showing the distribution with wavelength of the energy emitted by black-body sources at temperatures of 5750 K, approximating to that of the sun; and 245 K, a typical temperature for a planetary surface or a planetary atmosphere. Notice that the two curves are almost entirely separate. Also shown in fig. 2.1 is the absorption by various constituents in the earth's atmosphere; electronic bands occur in the ultraviolet and vibration–rotation and pure rotation bands due to minor constituents in

Fig. 2.1. (a) Curves of black-body energy B_λ at wavelength λ for 5750 K (approximating to the sun's temperature) and 245 K (approximating to the atmosphere's mean temperature). The curves have been drawn of equal areas since integrated over the earth's surface and all angles the solar and terrestrial fluxes are equal.

   (b) Absorption by atmospheric gases for a clear vertical column of atmosphere. The positions of the absorption bands of the main constituents are marked.

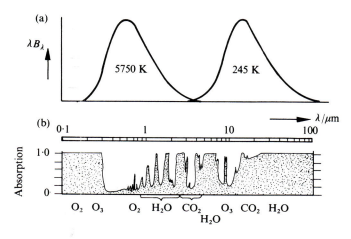

the infrared. The major constituents nitrogen and oxygen, because of their symmetry, possess no electric dipole transitions and hence no strong bands due to them occur in the infrared.

In § 1.4 an atmosphere was considered completely transparent to both solar and thermal radiation, in which heat was transferred from the heated surface by vertical convection. Inspection of fig. 2.1 suggests a slightly more elaborate model of an atmosphere transparent to solar radiation so that the surface is heated as before, but possessing an absorption coefficient uniform throughout the emitting infrared region and independent of pressure or temperature. Such an atmosphere is said to be *grey*.

The absorbing constituents in an atmosphere are molecules with complicated absorption spectra, much more complicated in fact than is indicated in fig. 2.1 (b) (cf. fig. 4.2). The approximation of a single absorption coefficient independent of wavelength, pressure and temperature is a very crude one. However, because of the large amount of overlap of the spectral lines involved, particularly in the lower atmosphere, the approximation is not, in fact, so crude as at first appears and it will enable us to derive features of the basic structure of an atmosphere in which radiative transfer is important. For a more accurate treatment and especially for conditions in the upper atmosphere, more detailed consideration of the spectral structure is necessary. This will be given in chapter 4.

2.2   Absorption and emission

First it is necessary to derive some basic equations in radiative transfer. The equations will be derived for absorption and emission processes only. In the absence of clouds, scattering is unimportant at the infrared wavelengths we are considering and is, therefore, ignored for the moment. Radiative transfer in clouds is treated in chapter 6; in § 6.4 it is shown that scattering processes may easily be incorporated into the same basic equations.

In a plane parallel atmosphere uniform in the horizontal consider first absorption of radiation along the vertical co-ordinate z. The law of absorption (sometimes known as Lambert's law or Bouguet's law) states that the absorption which occurs when radiation of intensity I (units: power per unit area per unit solid angle) traverses an elementary slab of atmosphere of thickness dz is proportional to the mass of absorber $\rho \, dz$ in unit cross-section of the slab (ρ being the density of absorber) and to the incident intensity of radiation itself. Thus

$$dI = -Ik\rho \, dz \tag{2.1}$$

where k is the absorption coefficient. Integrating (2.1) leads to

$$I = I_0 \exp\left(-\int k\rho \, dz\right) \tag{2.2}$$

The quantity $\exp(-\int k\rho \, dz)$ is the *fractional transmission* τ of the path, and $\int k\rho \, dz$ is known as the *optical path* χ (when measured from the top of the atmosphere downwards it is the *optical depth*).

A slab of atmosphere will also emit radiation in amounts depending on its temperature. If thermodynamic equilibrium applies, application of Kirchhoff's law enables the amount emitted per unit area to be written as $k\rho \, dz \, B(T)$ where $B(T)$ is the black-body emission per unit solid angle per unit area of a surface at temperature T.

From the Stefan–Boltzmann law, the integral of $B(T)$ over a hemisphere is proportional to T^4, i.e.

$$\int_{2\pi} B \, dS \cos \theta \, d\omega = \sigma T^4 \, dS$$

giving $B = \pi^{-1} \sigma T^4$

where σ is the Stefan–Boltzmann constant and $d\omega$ is an element of solid angle at an angle θ to the normal to the element of surface area dS.

The earth's atmosphere is not, of course, precisely in thermodynamic equilibrium; however, in the lower atmosphere conditions known as *local thermodynamic equilibrium* (LTE) prevail which constitute a sufficiently good approximation to equilibrium for black-body emission to be employed in the equation for emission. We shall come back to this point later when considering the high atmosphere where LTE no longer applies (§ 5.6).

The equation for radiative transfer through the slab, which includes both absorption and emission, is sometimes known as *Schwarzschild's equation*:

$$dI = -Ik\rho \, dz + Bk\rho \, dz$$

or $\quad \dfrac{dI}{d\chi} = I - B \tag{2.3}$

In deriving (2.3), vertically travelling radiation only has been considered. Radiation is, of course, travelling in all directions. However, under the assumption of a plane parallel atmosphere the problem is reduced to a one dimensional one by considering the two fluxes $F^\uparrow$ and $F^\downarrow$ which are the quantities $\int I(\theta) \cos \theta \, d\omega$ integrated over the downward facing and upward facing hemispheres respectively, $I(\theta)$ being the intensity at an angle θ to the vertical and $d\omega$ an element of solid angle (fig. 2.2). Detailed calculation shows that to a good approximation in (2.3) I may be replaced by F if dz is replaced by a mean thickness $\frac{5}{3}dz$ and B is replaced by πB, the black-body function integrated over a hemisphere.

2.3   Radiative equilibrium in a grey atmosphere

Having now laid the theoretical foundation of radiative transfer in a grey atmosphere, we apply it to find the equilibrium situation in an atmosphere in which transfer of infrared radiation is the only energy transfer mechanism and which has for its lower boundary a heated surface at temperature T_g.

Under these assumptions the net rate of temperature change, dT/dt, which occurs in a slab of plane parallel atmosphere due to the divergence of the upward and downward radiation fluxes is given by (fig. 2.3)

$$\frac{d}{dz}(F^{\downarrow}-F^{\uparrow})=\rho c_p\frac{dT}{dt} \tag{2.4}$$

In equilibrium $dT/dt=0$, and integration of (2.4) gives

$F^{\uparrow}-F^{\downarrow}=$ a constant ϕ, the net flux.   (2.5)

Further, from (2.3) the transfer equations are

$$\frac{dF^{\uparrow}}{d\chi^*}=F^{\uparrow}-\pi B$$

$$-\frac{dF^{\downarrow}}{d\chi^*}=F^{\downarrow}-\pi B \tag{2.6}$$

Fig. 2.2

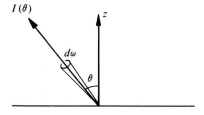

Fig. 2.3

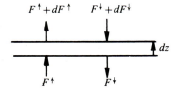

where the star on the optical depth χ^*, which is measured from the top of the atmosphere, denotes that it is appropriate to hemispheric radiation (i.e. the factor $\frac{5}{3}$ has been applied to dz in the expression for χ).

If

$$\psi = F^\uparrow + F^\downarrow \tag{2.7}$$

equations (2.6) may be written

$$\frac{d\psi}{d\chi^*} = \phi \tag{2.8}$$

$$\frac{d\phi}{d\chi^*} = \psi - 2\pi B \tag{2.9}$$

Since $\phi = $ constant (from (2.5)) $d\phi/d\chi^* = 0$

and

$$\psi = 2\pi B \tag{2.10}$$

which on substitution into (2.8) gives

$$B = \frac{\phi}{2\pi}\,\chi^* + \text{constant} \tag{2.11}$$

The boundary condition at the top of the atmosphere $(\chi^* = 0)$ is $F^\downarrow = 0$, so that here $\psi = \phi$ and, from (2.10), the constant in (2.11) is $\phi/2\pi$, i.e.

$$B = \frac{\phi}{2\pi}\,(\chi^* + 1) \tag{2.12}$$

At the bottom of the atmosphere where $\chi^* = \chi_0^*$, $F^\uparrow = \pi B_g$, B_g being the black-body function at the temperature of the ground. It is easy to show that there must be a temperature discontinuity at the lower boundary, the black-body function for the air close to the ground being B_0, and

$$B_g - B_0 = \frac{\phi}{2\pi} \tag{2.13}$$

The black-body function B for such an atmosphere is plotted against χ^* in fig. 2.4. The equivalent diagram in terms of temperature and height is fig. 2.5. Notice that the lower part of the atmosphere possesses a very steep lapse rate of temperature, and at the surface itself there is a discontinuity in temperature. As was shown in § 1.4, such a steep lapse rate is very unstable with respect to vertical motion, and will soon be destroyed by the process of *convection* which will tend to establish a mean adiabatic lapse rate.

In fig. 2.5 air from near the surface will tend to rise along the line which has been drawn with a slope of $-6\ \text{K km}^{-1}$ (a mean adiabatic lapse rate, see § 1.4 and also § 3.2) and which cuts the radiative equilibrium curve at the height of approximately 10 km.

This simple model does, in fact, bear some relation to what occurs in the real atmosphere. The *tropopause* is a surface situated at a height of ~ 10 km in mid latitudes which divides the region below (the *troposphere* or turning-sphere), in which convection is the dominant mechanism of vertical heat

Fig. 2.4. Upward radiation flux $F^\uparrow$, downward flux $F^\downarrow$ and black-body function πB at atmospheric temperature plotted against optical depth for radiative equilibrium atmosphere.

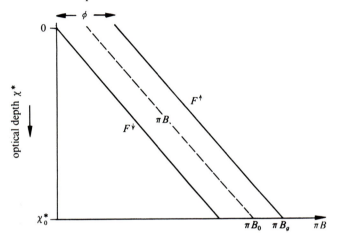

Fig. 2.5. Radiative equilibrium temperature T plotted against altitude z. The line c is drawn through the surface temperature with slope $-6 \, \text{K} \, \text{km}^{-1}$. This simple model leads to a troposphere dominated by convection below a stratosphere in approximate radiative equilibrium.

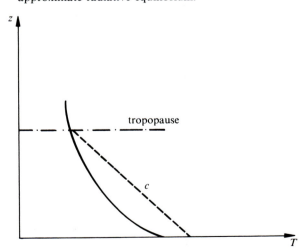

transfer, from the *stratosphere* which is much more stably stratified and where radiative transfer is dominant.

2.4 Radiative time constants

In considering radiative processes it is important to know the rate at which changes may be expected to occur. The time constant appropriate to any particular process will depend on the scale of the phenomenon under consideration, and the density of the air at the altitude in question, as well as on the details of the process itself. For the atmosphere as a whole a rough estimate may be made by considering a slab of atmosphere of thickness h and uniform density ρ radiating like a black body at a temperature $T \simeq 270$ K, different by a small amount ΔT from what it would be if in radiative equilibrium with the surrounding layers both above and below the slab.

Its rate of change of temperature is given by

$$c_p \rho h \frac{d \Delta T}{dt} = 8 \sigma T^3 \Delta T \qquad (2.14)$$

The time constant resulting from (2.14) is $c_p \rho h / 8 \sigma T^3$. If ρ is the density appropriate to the 50 kPa level and h equal to the scale height H, i.e. ~ 8 km, the magnitude of this time constant is about 6 days. Radiative processes in the lower atmosphere, therefore, in general, act rather slowly and for the consideration of short-term atmospheric developments may often be neglected. In the longer term, however, because they are the processes which largely determine the distribution of energy sources and sinks they are of dominant importance.

2.5 The greenhouse effect

Combining (2.12) and (2.13) we find that for the radiative equilibrium atmosphere:

$$B_g = \frac{\phi}{2\pi} (\chi_0^* + 2) \qquad (2.15)$$

where χ_0^* is the optical depth at the bottom of the atmosphere. If $\chi_0^* = 0$, $B_g = \phi/\pi$ and the surface temperature is in equilibrium with the incoming and the outgoing radiation, which are both equal to ϕ. If χ_0^* is large, the surface temperature represented by the black-body function B_g will be very considerably enhanced, an illustration of the so-called *greenhouse effect* mentioned in § 1.2.

In practice, the optical depth of an atmosphere will depend on the concentration of gases, such as water vapour, which are evaporated from the surface and whose concentration may increase rapidly with surface temperature. A mechanism for positive feedback therefore exists which may be particularly important in the case of the Venus atmosphere (Rasool & DeBergh, 1970; Ingersoll, 1969). It has been called the *runaway greenhouse effect* and is illustrated for three planets in fig. 2.6. Suppose the atmospheres of these planets began to form by outgassing from their interiors at a time when their surface temperatures were essentially determined by equilibrium between absorbed solar radiation and emitted long-wave radiation ((1.1) with $A \simeq 0$ because no clouds would be present) at values given by those on the left hand side of fig. 2.6. As water vapour accumulated in an atmosphere, owing to the greenhouse effect the surface temperature rose, so increasing the evaporation from the surface until either the atmosphere became saturated with water vapour or until all the available water had evaporated. The formation of clouds by condensation of water vapour enhances the effect except for the case of Mars where the atmosphere is so thin that significant cloud formation has not occurred and the blanketing effect of the atmosphere is small. For the earth, the equilibrium situation is with most of the water in liquid form, while for Venus, on these assumptions, the surface temperature would always be above the boiling point of water at the surface pressure. No liquid water would, therefore, be present on Venus. Supposing that originally similar amounts of water were present as for the earth, the early Venus atmosphere would have water vapour as the major constituent. No other gases would be present to prevent ultraviolet solar radiation dissociating water vapour at the top of the atmosphere; the

Fig. 2.6. Illustrating the greenhouse effect for the terrestrial planets. Their surface temperature is plotted against the vapour pressure of water vapour in the atmosphere. Also on the diagram (dashed) are the phase lines for water, the shaded area showing where liquid water is in equilibrium. For Mars and the Earth the greenhouse effect is halted when water vapour becomes saturated with respect to ice or water. For Venus the diagram illustrates the 'runaway greenhouse effect'. (After Rasool & De Bergh, 1970 and Goody & Walker, 1972)

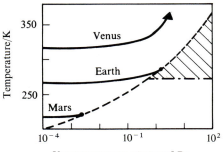

resulting hydrogen would escape (§ 5.3) and the oxygen would be consumed in various oxidation processes at the surface. An explanation is, therefore, afforded of the small quantity of water on Venus today compared with the large amount on the earth. The large amount of carbon dioxide remaining in the Venus atmosphere rather than as carbonates in the rocks is also consistent with this atmospheric history.

Problems

2.1 If one third of the solar energy incident outside the atmosphere were absorbed by the atmosphere, show that the average rate of temperature rise (assuming no loss) of the atmosphere would be $\sim 1\,K$ per day. Rates of temperature change due to radiative processes are often expressed in units of K per day. For instance an average rate of loss of energy by the troposphere due to emission of long-wave radiation is $\sim 1\,K$ per day.

2.2 Assuming an absorber with uniform absorption coefficient and uniformly mixed with height, a surface temperature of 280 K, a tropopause temperature at 25 kPa pressure of 220 K and radiative equilibrium conditions, find ϕ. Also find χ^* in equation (2.12) as a function of pressure. What would be the temperature discontinuity at the surface under such assumptions?

2.3 The calculation of § 2.3 has been carried out on the assumption that radiative equilibrium exists throughout the atmosphere. Make a qualitative assessment of the effect on the radiative equilibrium profile above the tropopause of a change in the profile below the tropopause from the radiative equilibrium one to the convective one (fig. 2.5).

2.4 Estimate the radiative time constant for the atmosphere of Mars.

2.5 What is the length of a solar day on Venus? (cf. table 1.2 noting also that the sense of rotation of Venus is opposite to that of the earth and Mars). Ignoring the effect of the clouds, by a similar calculation to problem 2.4, estimate the radiative time constant for the lower Venus atmosphere. Because of the blanketing effect of the clouds (fig. 2.7) and probably also because of motion in the Venus atmosphere, no difference in temperature in the lower Venus atmosphere has been measured between the day and the night side.

2.6 The surface temperature of Venus is $\sim 750\,K$ (table 1.1). From the value of T_m from table 1.2 and from (2.15) assuming radiative equilibrium, calculate the optical depth of the Venus atmosphere.

Fig. 2.7. An idea of the variability of cloud features on Venus can be obtained from these four images taken in the ultra-violet from the Pioneer Venus orbiter over a period of 38 hours in May 1980. Note the polar rings which are brighter than the other cloud features because of the greater number at these latitudes of small particles which scatter ultra-violet light more effectively (cf. § 4.1) (from Briggs & Taylor 1982)).

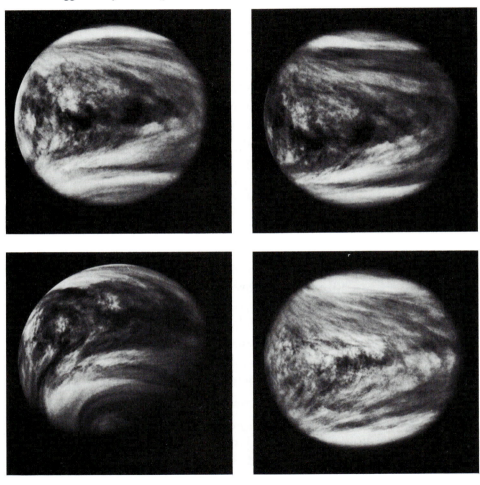

3

Thermodynamics

3.1 Entropy of dry air

In chapter 1 the first law of thermodynamics was applied to vertical motions in a planetary atmosphere and the adiabatic lapse rate for dry air was derived. On substituting $T\,dS$ for dq in (1.7), where S is the entropy, and substituting for V from the equation of state for a perfect gas we have

$$dS = c_p \frac{dT}{T} - \frac{R\,dp}{M_r\,p} \tag{3.1}$$

which on integration gives an expression for the entropy per unit mass:

$$S = c_p \ln T - RM_r^{-1} \ln p + \text{constant} \tag{3.2}$$

If air at pressure p and temperature T is brought adiabatically (i.e. at constant S) to a standard pressure p_0 of 100 kPa, its temperature θ at that standard pressure is known as its *potential temperature*.

From (3.2)

$$c_p \ln \theta = c_p \ln T - RM_r^{-1} \ln p + RM_r^{-1} \ln p_0 \tag{3.3}$$

from which, remembering that for a perfect gas $c_p - c_v = R/M_r$,

$$\theta = T \left(\frac{p_0}{p} \right)^\kappa \tag{3.4}$$

where $(c_p - c_v)/c_p = \kappa = 0.288$ for dry air in the earth's atmosphere.

Also

$$S = c_p \ln \theta + \text{a constant} \tag{3.5}$$

– a useful equation because it enables us to express the entropy of dry air in terms of the more readily interpretable concept of potential temperature.

3.2 Vertical motion of saturated air

The structure of an atmosphere may be influenced very considerably by the presence within it of a condensable vapour. In the case of the earth's atmosphere water vapour is very important.

Consider the vertical motion of air containing water vapour. Provided air remains unsaturated, its thermodynamic properties are little different from those of dry air (problem 3.1). As unsaturated air rises, therefore, it cools similarly to dry air at about 10 K km^{-1}. During this ascent although the *mixing ratio*† of water vapour remains constant, the *relative humidity*‡ increases and may reach 100%. The level at which this occurs is the *condensation level*. As the air continues to rise, it remains saturated, the surplus water vapour condensing to form drops of liquid water. The latent heat released by this condensation process must now be included.

The same assumptions are made as in § 1.4, together with the additional assumption that the parcel contains liquid water which moves with it; the air remains saturated, evaporation and condensation occurring so that equilibrium is continually maintained. Consider a parcel of 1 g of dry air with m g of water vapour (m is the *saturation mixing ratio*) and $\xi - m$ g of liquid water. For the dry air the entropy is given by (3.2). The entropy of the water vapour and liquid water is the entropy of ξ g of liquid water at temperature T plus the additional entropy Lm/T required to convert m g of liquid water to water vapour. The total entropy S is therefore

$$S = (c_p + \xi c) \ln T - \frac{R}{M_{ra}} \ln (p - e) + \frac{Lm}{T} + \text{constant} \tag{3.6}$$

where c is the specific heat of liquid water, e the saturation vapour pressure and L is the latent heat which, of course, varies with temperature. Differentiating (3.6) and substituting from the hydrostatic equation which for wet air is

$$dp = -g\rho_a(1 + \xi)\, dz \tag{3.7}$$

we have, for adiabatic conditions,

$$(c_p + \xi c)\frac{dT}{T} + d\left(\frac{Lm}{T}\right) + \frac{R}{M_{ra}(p - e)}\cdot\frac{de}{dT}\, dT + \frac{g}{T}(1 + \xi)\, dz = 0 \tag{3.8}$$

The equations of state for unit masses of the air (density ρ_a, molecular weight M_{ra}) and water vapour (density ρ_v, molecular weight M_{rv}) considered separately are

$$p - e = \rho_a RT/M_{ra} \tag{3.9}$$

---

† i.e. the ratio of number of water molecules to other molecules in a given volume (the volume mixing ratio) or the ratio of the mass of water vapour to mass of air in a given volume (the mass mixing ratio).

‡ the ratio of water vapour pressure to saturation water vapour pressure at air temperature.

and $\qquad e = \rho_v RT/M_{rv}$ (3.10)

Now the mixing ratio m is dependent on air pressure as well as temperature

$$m = \frac{\rho_v}{\rho_a} = \left(\frac{e}{p-e}\right)\frac{M_{rv}}{M_{ra}} \simeq \frac{e\varepsilon}{p}$$ (3.11)

since, in the lower atmosphere, $e \ll p$, and

$\varepsilon = M_{rv}/M_{ra} = 0.622$

Differentiating (3.11) gives

$$\frac{dm}{dz} = \frac{\varepsilon}{p}\frac{de}{dT}\cdot\frac{dT}{dz} - \frac{\varepsilon e}{p^2}\cdot\frac{dp}{dz}$$ (3.12)

Substituting from (3.12) in (3.8), considering the air just saturated, i.e. $\xi = m$, and substituting from the Clausius Clapeyron equation

$$\frac{de}{dT} = \frac{LeM_{rv}}{RT^2}$$ (3.13)

we have

$$-\frac{dT}{dz} = \Gamma_s$$

$$= \frac{g}{c_p}\cdot\frac{\{1+(LeM_{rv}/pRT)\}\{1+(e\varepsilon/p)\}}{[1+(\varepsilon e/pc_p)\{c+(dL/dT)\}+(\varepsilon eL^2 M_{rv}/c_p pRT^2)]}$$ (3.14)

On inserting typical values (problem 3.2) the middle term in the denominator on the right hand side turns out to be negligible in comparison with the other terms. Also, since $e \ll p$, the second bracket in the numerator $\simeq 1$, so that (3.14) becomes

$$\Gamma_s = \Gamma_d\left(1+\frac{LeM_{rv}}{pRT}\right)\left(1+\frac{LeM_{rv}}{pRT}\cdot\frac{\varepsilon L}{c_p T}\right)^{-1}$$ (3.15)

Γ_s is always $< \Gamma_d$ and varies in the atmosphere from about $0.3\Gamma_d$ to Γ_d. Actual values can be read from fig. 3.1.

The stability of saturated air containing liquid water for vertical motion can be considered in exactly the same way as that for dry air in § 1.4. We have stable or unstable conditions according as $-dT/dz$ is $<$ or $>\Gamma_s$. It should be noted that for saturated air with no liquid water content the stability condition for upward motion involves Γ_s, while that for downward motion involves Γ_d, since as soon as the air moves downward it becomes unsaturated.

The assumption made above, that the liquid water remains with the parcel as it is lifted, is clearly not necessarily fulfilled in practice. If, however, it is assumed that the products of condensation, or a proportion of them, fall out during the ascent, little difference is made to the final answer. Another assumption which has been made in considering vertical motion is that the environment is unmodified by the rising or sinking parcels of air. The motion of a parcel of air is, of course, bound to produce a compensating motion in the environment. However, since the compensating descending motions normally occur over a much larger area than the ascending motion, the *parcel* method we have described, which ignores the compensating motions, is in general a good approximation and is used in general forecasting analysis.

3.3   The tephigram

The tephigram is a thermodynamic diagram particularly suitable for representing atmospheric processes. The name derives from T–ϕ-gram since it is a temperature–entropy plot, the symbol ϕ sometimes being used for entropy (we have used the symbol S). Instead of plotting S as the ordinate in our thermodynamic diagram, we can plot $\ln \theta$ (3.5) remembering that θ is the potential temperature referred to *dry* air. The basic diagram therefore is a plot of $\ln \theta$ against T. Contours of other quantities may also be plotted on the tephigram to assist in its use, namely:

(1) The pressure p, from (3.4), is a function of T and θ, so isobars or lines of constant pressure can be drawn. For many applications it is then convenient to use the diagram as one in which T can be plotted as a function of p.

(2) Adiabatic lines for dry air are lines of constant θ.

(3) Adiabatic lines for saturated air may be plotted from (3.15) where the lapse rate is given as a function of T and p. These *wet adiabatics* show the entropy change of the saturated air which is exactly equal but opposite in sign to that of the liquid water carried with it (or falling out of it).

(4) The saturation mixing ratio m from (3.11) is a function of T and p only. Note that a plot of mixing ratio against pressure for unsaturated air, is identical to a plot of *dew point* (or frost point for temperatures below zero celsius) against pressure).

Fig. 3.1 is an example of a tephigram on which the above isopleths are drawn. Problems at the end of the chapter illustrate its use in some detail. Also figs. 3.4, 3.5 and 3.6 illustrate from satellite and radar pictures some of the thermodynamic processes involved in the atmosphere.

3.4   Total potential energy of an air column

The energy dE of an element of a static air column at altitude z consists of two parts, its internal energy dE_I which is $c_v T$ per unit mass and its potential energy dE_P which is gz per unit mass. Since an element of mass is $-dp/g$ (1.2) we

Fig. 3.1. Part of a tephigram. (Charts available in the UK from Her Majesty's Stationery Office.)

  The main axes of entropy and temperature are inclined. Isopleths of pressure (marked in mb, 1000 mb = 100 kPa) are then nearly parallel to the lower edge of the page.

  The dashed lines are lines of constant water vapour mixing ratio (labelled in g kg^{-1}) and the most curved lines are wet adiabatics.

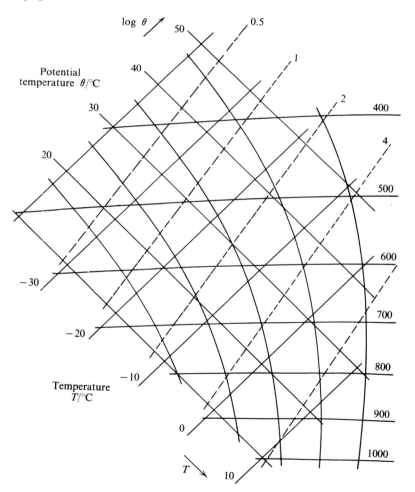

have

$$E = E_I + E_P = g^{-1} \int_{p_h}^{p_0} c_v T \, dp + \int_{p_h}^{p_0} z \, dp \tag{3.16}$$

where the column extends from $z = 0$, $p = p_0$ to $z = h$, $p = p_h$. The second term on the right hand side may be integrated by parts to give

$$E_P = -p_h h + \int_0^h p \, dz \tag{3.17}$$

Now since $p = R\rho T / M_r$ (equation of state (1.3))

and $\rho \, dz = -dp/g$, we have

$$E_P = -p_h h + g^{-1} \int_{p_h}^{p_0} \frac{RT}{M_r} \, dp \tag{3.18}$$

The first term may be made negligible by choosing h sufficiently high that $p_h \simeq 0$. Substituting from (3.18) into (3.16), remembering that $c_p = c_v + R/M_r$, we have

$$E = E_I + E_P = c_p g^{-1} \int_0^{p_0} T \, dp \tag{3.19}$$

Note that $E_P / E_I = c_p / c_v - 1 \simeq 0.4$ for dry air so that, to the extent that hydrostatic equilibrium prevails, the potential and internal energies of a column of air bear a constant ratio to each other. Conversion into kinetic energy will, therefore, occur at the expense of both potential and internal energy, in this same ratio. It is, therefore, convenient to treat potential and internal energy together; their sum is called *total potential energy* – a name first used by Margules in a famous paper in 1903 on the energy of storms.

3.5   Available potential energy

All the total potential energy is clearly not available for conversion into kinetic energy. For instance, for a uniformly stratified atmosphere, stable with respect to vertical motion, in which there are no variations of density in the horizontal at any level, although the total potential energy is large, none at all is available for conversion. Suppose now that such a stratified atmosphere is heated in a restricted region. Total potential energy is added and the stratification is disturbed; horizontal density gradients and hence pressure forces are created which may convert total potential energy into kinetic energy. Suppose, further, that a uniformly stratified atmosphere is cooled over a limited region rather than heated. Although total potential energy is removed, as a result of this removal the stratification is disturbed and conversion into kinetic energy is again possible. Removal of energy can be as effective as addition of energy in making more energy available.

A quantity which for any given atmospheric state is a measure of the total potential energy which is available for conversion through adiabatic processes into kinetic energy was introduced by Lorenz (1955) and is called *available potential energy*. It is the difference between the total potential energy in the state under consideration, and the total potential energy of a state of uniform stratification which is statically stable (i.e. having a stable lapse rate of temperature) obtained by a redistribution of atmospheric mass by adiabatic processes. For conditions of such uniform stratification the available potential energy is zero.

The process of readjustment to the statically stable reference state involves the imposition of vertical motions so that the originally undulating isentropic surfaces (i.e. surfaces of constant potential temperature) are brought into coincidence with surfaces of constant geopotential (cf. § 7.5 and problem 7.11 for definition of geopotential). Since warmer air rises and cooler air descends during the readjustment, gravitational potential energy will be released and work will be done at the expense of internal energy; the sum of these is the available potential energy we wish to find.

To calculate its magnitude, following Lorenz (1955), consider the atmosphere divided up by surfaces of constant potential temperature θ (§ 3.1), such as those illustrated by the example of fig. 3.2.

In any redistribution under adiabatic flow the value of θ for any given element of atmosphere will be conserved. Provided that each of the surfaces of constant θ intersect every vertical column only once, we can easily define the average pressure $\bar{p}$ over each of the surfaces with respect to the area projected on to a horizontal surface. Surfaces which appear to intersect the ground may be imagined to continue at ground level, i.e. where $p = p_0$. The total potential energy E of a vertical column (3.19) can be written in terms of potential temperature θ from (3.4) and integrated by parts to give

$$E = (1+\kappa)^{-1} c_p g^{-1} p_0^{-\kappa} \left(\int_{\theta_0}^{\infty} p^{1+\kappa} \, d\theta + \theta_0 \, p_0^{1+\kappa} \right) \qquad (3.20)$$

Provided that the value of θ at the surface, namely θ_0, is lower than any value in the atmosphere above (this is a requirement for stability; see also fig. 3.2), the last term on the right hand side of (3.20) can be incorporated in the integral by replacing the lower limit with $\theta = 0$ with the interpretation that all surfaces of θ with values less than θ_0 coincide at $p = p_0$. We therefore have for the whole atmosphere (per unit area of surface)

$$E = (1+\kappa)^{-1} c_p g^{-1} p_0^{-\kappa} \int_0^\infty \overline{p^{1+\kappa}} \, d\theta \qquad (3.21)$$

where $\overline{p^{1+\kappa}}$ is the average of $p^{1+\kappa}$ over an isentropic surface, the average to be taken in the same way as when defining the average pressure $\bar{p}$.

Fig. 3.2. (a) Average temperature cross-section of atmosphere for northern hemisphere winter.

(b) Distribution of potential temperature θ corresponding to temperature distribution in (a).

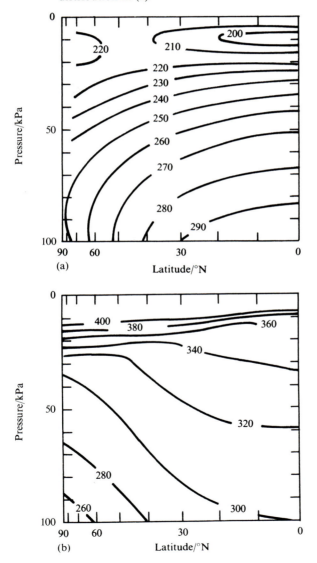

The minimum total potential energy which can result from adiabatic rearrangement occurs when $p = \bar{p}$ everywhere over a given isentropic surface. The difference between this minimum value of total potential energy and the original value is the *available potential energy* E_A and is, therefore, given by

$$E_A = (1+\kappa)^{-1} c_p g^{-1} p_0^{-\kappa} \int_0^\infty (\overline{p^{1+\kappa}} - \bar{p}^{1+\kappa}) \, d\theta \qquad (3.22)$$

Because $1+\kappa > 1$, $\overline{p^{1+\kappa}} > \bar{p}^{1+\kappa}$ unless $p \equiv \bar{p}$, so that (3.22) ensures that E_A is positive, as is, of course, necessary.

In (3.22) let $p = \bar{p} + p'$, $p^{1+\kappa}$ can then be expanded by the binomial theorem. Remembering that $\overline{p'} = 0$ and expanding to the first term only,

$$E_A = \tfrac{1}{2} \kappa c_p g^{-1} p_0^{-\kappa} \int_0^\infty \bar{p}^{1+\kappa} \overline{\left(\frac{p'}{\bar{p}}\right)^2} \, d\theta \qquad (3.23)$$

Equation (3.23) expresses E_A in terms of the variance of p on an isentropic surface. It is often more convenient to refer to isobaric surfaces, in which case an appropriate expression for E_A is (problem 3.12)

$$E_A = \tfrac{1}{2} c_p g^{-1} \int_0^\infty \bar{T} \left(1 - \frac{\Gamma}{\Gamma_d}\right)^{-1} \overline{\left(\frac{T'}{\bar{T}}\right)^2} \, dp \qquad (3.24)$$

Equation (3.24) is a useful form for estimating E_A. By comparing it, for instance,

(c) Potential temperature distribution which results when atmosphere shown in (b) is moved isentropically to a state of uniform horizontal stratification. (After Lorenz, 1967)

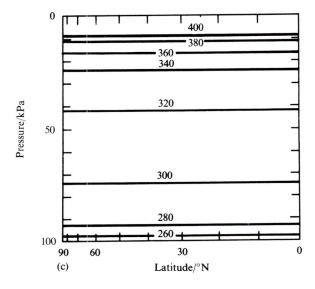

(c)

with (3.19), an estimate can be made of the ratio of available potential energy E_A to total potential energy E. Typical values in (3.24) are $\Gamma \simeq \frac{2}{3}\Gamma_d$ and $\overline{(T'/\bar{T})^2} \simeq (\frac{1}{16})^2$ giving $E_A/E \simeq \frac{1}{200}$. Only about $\frac{1}{2}\%$ of the total potential energy, therefore, is available for conversion into kinetic energy.

It is interesting to compare the average kinetic energy with both the total potential energy and the available potential energy. The total potential energy of a vertical column (3.19) may be written as

$$E = \{(\gamma-1)g\}^{-1} \int_0^{p_0} c^2 \, dp \tag{3.25}$$

where $\gamma = c_p/c_v$ and $c = (\gamma RT)^{1/2}$ is the speed of sound in air. The kinetic energy E_K of a vertical column expressed in a similar form is

$$E_K = \frac{1}{2g} \int_0^{p_0} V^2 \, dp \tag{3.26}$$

where V is the velocity of atmospheric motion. A typical value of V is $\sim 0.05c$ so that the ratio E/E_K may be ~ 2000 on average.

Since $E/E_A \simeq 200$, $E_K/E_A \simeq 0.1$ on average, so that the amount of available potential energy in the atmosphere is much larger than the amount of kinetic energy.

The concept of available potential energy may be employed to investigate the efficiency of heating or cooling different parts of the atmosphere in contributing to the maintenance of its general circulation. If the quantity $\dot{q} \, dm$ (dm being an element of mass) represents the rate at which heat is introduced or removed by diabatic processes (e.g. radiation, latent heat, turbulent transfer from the surface etc.), it can be shown that (see, for instance, Dutton & Johnson (1967)) the rate of generation of available potential energy is approximately

$$\int \left\{1 - \left(\frac{\bar{p}}{p}\right)^\kappa\right\} \dot{q} \, dm \tag{3.27}$$

Lorenz called the quantity $\{1 - (\bar{p}/p)^\kappa\}$ the efficiency factor. Inspection of fig. 3.2 (problem 3.14) shows that maximum generation of available potential energy will occur with low altitude heating in low latitudes and high altitude cooling in polar regions – a result which might be expected from the very general thermodynamic arguments of § 1.5.

In the above discussion the available potential energy has been computed by reference to a *statically stable* reference state. Since there are dynamical constraints on the motions which may occur in the atmosphere (see chapters 7ff.) it is not clear that processes for attaining such a reference state are dynamically possible. In fact, as has been suggested by Van Mieghem (see Dutton & Johnson, 1967), it may be more realistic to postulate a reference state which includes a circumpolar vortex balanced by a horizontal temperature gradient which may be reached by isentropic flows in the manner described above, but which also preserves the absolute zonal angular momentum.

3.6   Zonal and eddy energy

The above discussion has not been concerned at all with the way in which the conversion from potential energy to kinetic energy takes place. In chapter 10, two mechanisms for such conversion will be described, namely (1) through mean motions which may be considered independent of longitude, and (2) through large scale quasi-horizontal eddy motions which show large variations with longitude. For studies of the general circulation, therefore, it is convenient to divide the kinetic energy into two components by resolving atmospheric motion into (1) the mean zonal component, i.e. the wind velocity averaged around latitude circles, and (2) the eddy components, i.e. the

Fig. 3.3. Average storage and conversions of available potential energy E_A and kinetic energy E_K for the whole atmosphere as estimated by Oort & Peixoto (1974). Units of energy in 10^5 J m^{-2} and of conversion, generation or dissipation in W m^{-2}. Subscripts Z and E are for zonal and eddy components respectively.

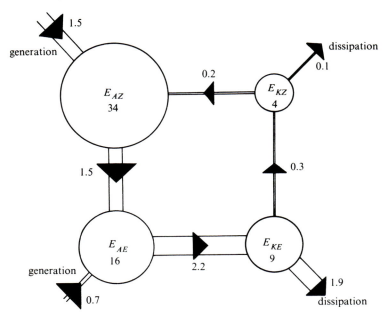

meridional (or north–south) component and the deviation of the zonal component at any place from the mean around the latitude circle. These components are known as the zonal kinetic energy and eddy kinetic energy respectively.

In a similar way zonal and eddy available potential energy may be defined. Components of the variance of the temperature field are chosen, namely (1) the variance of the zonally averaged temperature, and (2) the variance of temperature along latitude circles, from which zonal and eddy components of available potential energy may be deduced.

Fig. 3.3 shows an example of the result of a study of energy generation and conversion. Notice that most of the conversion to kinetic energy occurs from the eddy available potential energy which in turn is converted through large scale eddy motions from zonal available potential energy.

Problems

3.1 To allow in a convenient way for the different molecular weight M_{rw} which applies to air containing water vapour, a *virtual temperature* T^* is defined such that

$$RT^*/M_{ra} = RT/M_{rw}$$

where M_{ra} is the molecular weight of dry air. Derive an expression for the virtual temperature in terms of water vapour mixing ratio and find $T^* - T$ for saturated air at 100 kPa and 273 K, 290 K and 300 K.

3.2 Calculate typical values of the middle term in the denominator of (3.14) and show that for all atmospheric conditions, it is $\ll 1$. [dL/dT is approximately $-2\,\mathrm{J\,g^{-1}\,K^{-1}}$.]

3.3 Calculate values of the term $\varepsilon L/c_p T$ in (3.15) and show that because it is always > 1, $\Gamma_s < \Gamma_d$.

3.4 From the hydrostatic equation (1.2) show that the pressure $p(z)$ at height z is related to the field of potential temperature $\theta(z')$ by the equation

$$\left\{\frac{p(z)}{p(0)}\right\}^\kappa = \frac{\kappa g M_r}{R} \int_z^\infty \frac{dz'}{\theta(z')}$$

where $\kappa = (\gamma - 1)/\gamma$.

3.5 The following measurements were made at 0001 GMT on a day in June by a radiosonde ascent from Liverpool:

Pressure (kPa)	Temperature (°C)	Dew (or frost) point (°C)
100	13	11
94	9.5	8
90	7	5
78	0	− 3
70	− 5	− 11
60	− 11	− 17
50	− 20	− 28
40	− 32	− 42
30	− 47	
20	− 49	
15	− 50	
10	− 48	

Plot these on a tephigram chart and answer the following questions:

(1)   What is the pressure at the tropopause?
(2)   What parts of the ascent are stable for (a) dry air, (b) saturated air?
(3)   What is the mixing ratio, water vapour:air, at 100 kPa, 50 kPa?
(4)   If the surface cools radiatively during the night, how many degrees of cooling are required for fog to begin to form?
(5)   If air at the surface is heated during the following day and then rises adiabatically, at what level will condensation occur?
(6)   What will be the level of the top of convective clouds which develop during the day?

3.6 Air represented by the measurements given in problem 3.5 is forced to ascend either over a range of mountains or over a wedge of colder air at a cold front. Consider the layer of air initially between 78 and 70 kPa; note that it is stable for both dry and wet ascent. Suppose this air is forced to rise by 10 kPa in pressure. Note that it will remain 8 kPa deep in pressure units. Using a tephigram chart find the new temperature of the bottom of the ascent by taking it up the dry adiabatic until it becomes saturated, then up the wet adiabatic. Do the same for the top of the ascent. Hence show that the layer is now unstable with respect to wet ascent. Such a situation is called one of *potential instability*; the amount of such instability will determine how much convective development will occur.

Carry out the same exercise for the layer between 90 kPa and 78 kPa.

3.7 Again considering the ascent in problem 3.5, take air from the surface up the dry adiabatic to its condensation level, then up the wet adiabatic until it reaches the temperature of the environment. By measuring the appropriate area on the diagram calculate the work required to raise 1 kg of air to this point.

Now continue the ascent along the wet adiabatic until the parcel again reaches the temperature of the environment. Calculate the energy released in this second ascent.

Show that for the complete ascent the net energy release is positive. Such a condition is known as *latent instability*. Once the surface has been heated sufficiently to initiate vertical motion, energy can be released and intense convective activity and maybe thunderstorms are likely to develop (cf. fig. 3.6).

3.8 Two parcels of damp air at the same pressure with masses M_1 and M_2, water vapour mixing ratios m_1 and m_2, and temperatures T_1 and T_2 are mixed. Show that the resulting temperature and mixing ratio are given by

$$\bar{T} = \frac{M_1 T_1 + M_2 T_2}{M_1 + M_2}$$

$$\bar{m} = \frac{M_1 m_1 + M_2 m_2}{M_1 + M_2}$$

(remember, $m \ll 1$).

Using data available on the tephigram chart consider the mixing of two equal masses at 100 kPa pressure and 24°C and 12°C respectively. What relative humidity (assumed the same for each mass) must the masses possess for condensation to occur on mixing the masses together.

3.9 (1) Consider the vertical mixing of a column of air extending from pressure p_1 to pressure p_2. Show that the mixing ratio $\bar{m}$ of the mixed column is given by

$$\bar{m} = \frac{1}{p_2 - p_1} \int_{p_1}^{p_2} m \, dp$$

where $m(p)$ describes the initial distribution.

Show also with appropriate assumptions (state what they are)

Fig. 3. 4. A visible image from the NOAA-8 satellite taken at 0919 GMT 25 April 1984 and received and processed by the University of Dundee Electronics Laboratory. Sea fog, which had formed over the eastern Atlantic as warm air moved over cooler water, is seen passing to the north of Scotland and entering the Moray Firth as winds turn to a more northerly point. The fog, although originally warmer than the sea, cooled by long wave radiation until it became colder than the underlying sea surface. A vortex and lee waves (cf. § 8.3) which formed downwind of the Faeroe Islands are detectable in the structure of the fog. Upper cloud associated with a high-level cold pool casts shadows on the fog top. Snow patches remain on the highest points of the Scottish highlands.

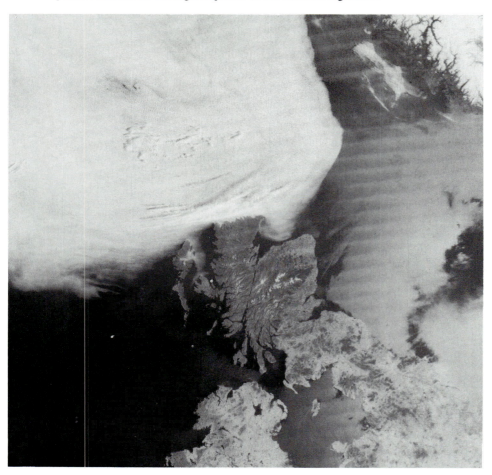

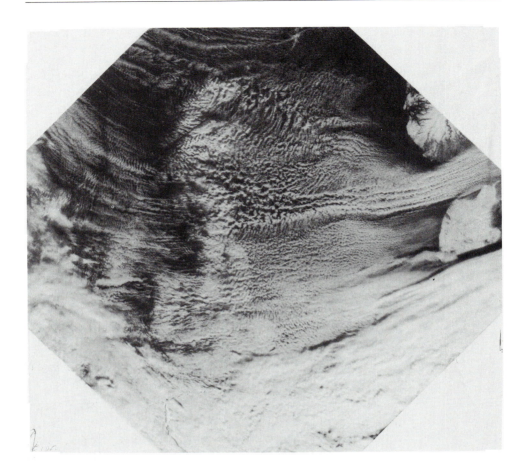

Fig. 3.5. Visible imagery obtained from a NOAA satellite at 1011 GMT on 14 February 1979 showing cloud lines associated with very cold continental air streaming westwards from Russia across the North Sea and British Isles. Parts of Norway and Denmark are visible in the east, while the British Isles can just be discerned beneath low-level convective cloud. With air temperatures over Denmark around $-12°C$ and sea temperatures around $5°C$, cumulus clouds formed rapidly over the sea in the gale-force Easterly winds. However, note the delayed formation of cloud to the lee of Norway where the air has been forced to descend. The depth of convection gradually increases as the air at low-levels becomes warmer and moister, giving rise to larger cumulus cells within the bands and the formation of showers. The earlier formation of cloud cells over the Skagerrak (between Norway and Denmark) can be seen to produce a strip of more mature cumulus which may be traced right across the North Sea. Over the eastern half of Britain, where the air temperature was around $-2°C$, snow showers were widespread. Amounts of precipitation were small, however, because the cloud tops extended up to only 2000 m. To the west of the high ground of central Britain, the

that an approximate expression for the potential temperature $\bar{\theta}$ of the mixed column is

$$\bar{\theta} = \frac{1}{p_2 - p_1} \int_{p_1}^{p_2} \theta \, dp$$

(2) Plot the following on a tephigram chart: 100 kPa, 28°C; 95 kPa, 24°C; 90 kPa, 21°C. The water vapour mixing ratios are 20 g kg^{-1} from 100 kPa to 95 kPa and 17 g kg^{-1} from 95 kPa to 90 kPa. If the 100 kPa to 90 kPa layer is mixed, estimate the resulting potential temperature and the condensation level. Your result demonstrates that mixing processes can lead to cloud or fog formation.

3.10 Plot on a tephigram chart the following large scale circulation cell between the equator and 30°N. Air rises adiabatically from 100 kPa and 25°C to 80 kPa when it becomes saturated and continues along the saturated adiabatic to 18.5 kPa and −75°C. It then moves northwards and sinks losing heat radiatively to 28 kPa and −72°C after which it descends adiabatically to the surface and returns along the surface while being heated to its starting point. Calculate the energy released in the cycle by unit mass of air.

    If the overturning takes 6 months compare the rate of release of energy with the average input of solar radiation.

3.11 From the information contained in fig. 3.2 carry out a numerical integration of (3.22) and find the available potential energy in the northern hemisphere. Note that since fig. 3.2 contains zonally averaged information your answer refers to zonal available potential energy. Compare your result with that in fig. 3.3.

3.12 Derive (3.24) from (3.23) as follows:

  (1)   If on an isobaric surface $\theta = \bar{\theta} + \theta'$, $\bar{\theta}$ being an average value over the surface, show that on a neighbouring isentropic surface where $p = \bar{p} + p'$,

$$p' = \theta' \, dp/d\theta$$

air dried out and much less convection is evident, although some snow showers occurred over Ireland. The main feature in the west is the large area of transverse lee waves (cf. §8.3) in the stratocumulus cloud which has formed beneath a marked inversion at 2 km height.

    Around the southern portion of this image there are extensive layers of upper cloud associated with an old frontal system. Shadows can be seen along the northern edges of these layers, particularly across southern Denmark. (Courtesy of the Meteorological Office)

(2)   From (3.4) show that

$$\frac{d\theta}{dp} = -\kappa\theta p^{-1}\left(1-\frac{\Gamma}{\Gamma_d}\right)$$

Fig. 3.6. (a) The cloud image (AVHRR visible channel received by University of Dundee Electronics Laboratory) was taken from the TIROS N satellite at 1428 GMT on 26 June 1980. The cloud seen over central and southern Britain and the near Continent is typical of that observed in a polar airmass on a summer's afternoon. Cold air passing over land warmed by the sun becomes unstable, generating updraughts and cloud. Over the neighbouring cooler sea, convection is largely absent.

Many of the clouds are bright, and therefore thick (cf. § 6.4) with sharply defined edges; some can be seen to cast shadows on their north-eastern side. Some of the cloud systems extend over some tens of kilometres, while others are very small either because they are still growing or because their growth is inhibited by the compensating descent of air surrounding the vigorous updraughts of the larger systems.

where $\Gamma = -dT/dz$ is the local lapse rate of temperature and $\Gamma_d = g/c_p$ is the adiabatic lapse rate.

(3) Make the assumption that in the result of part (1) the average vertical stability may be employed, i.e. $p' = -\theta' \, d\bar{p}/d\theta$, and substitute from parts (1) and (2) in (3.23) to derive (3.24).

(b) The radar network picture (courtesy of the Meteorological Office) was obtained at a time corresponding to the satellite picture (radar techniques for precipitation measurement are described in § 12.2). Many of the features seen on the satellite imagery can be seen to be correlated with the echoes from rainfall detectable by the radars. Note in particular the rain associated with the cloud systems just west of the Wash, over South Wales, around the Severn Estuary and over the south coast. Some of the rain was heavy, as depicted by white areas in the radar picture, and thunderstorms were reported.

In the early stages of growth, the clouds consist of supercooled water droplets which freeze (glaciate) as the air rises. When the ice particles grow to a size such that their terminal velocities exceed the updraught velocity in the cloud, they fall to the ground, sweeping up supercooled droplets lower in the cloud. The smaller ice particles continue to be carried in the updraught, reach the upper troposphere and are blown downwind. These are responsible for characteristic 'anvil clouds' on a showery day and can be detected on the satellite picture as wispy cloud to the south-east of some of the bright echoes.

3.13 Compare the average input of solar radiation over the hemisphere with the following quantities in fig. 3.3:

(1)   rate of generation of zonal available potential energy,

(2)   the rate of generation of kinetic energy.

3.14 From the information in fig. 3.2 plot a cross-section of the efficiency factor for generation of available potential energy (see (3.27)).

4

More complex radiation transfer

4.1   Solar radiation: its modification by scattering

In discussing radiation transfer in chapter 2 a very simplified atmospheric model was presented. The assumptions were made of an atmosphere in radiative equilibrium, transparent to solar radiation and possessing an absorption coefficient for long-wave radiation independent of frequency. In this chapter we shall consider how solar radiation is modified by the atmosphere and then set up the radiative transfer problem more generally showing how, for any given atmosphere, integrations over height and over frequency can be carried out.

Solar radiation is first modified by scattering processes within the atmosphere. A scattering coefficient σ can be defined in a similar way to the absorption coefficient k in equations (2.1) and (2.2) such that the transmission along an atmospheric path is given by

$$I = I_0 \exp\left(-\int \sigma\rho \, dz\right) \tag{4.1}$$

Apart from scattering by clouds, which will be dealt with in chapter 6, the most important scattering mechanism is Rayleigh scattering by molecules, for which the scattering coefficient σ_R at wavelength λ is given by

$$\sigma_R = \frac{32\pi^3}{3N_0\lambda^4\rho_0}(n-1)^2 \tag{4.2}$$

where N_0 is the number of molecules per unit volume, ρ_0 the density and n the refractive index, all at conditions of standard temperature and pressure. Attenuation by Rayleigh scattering varies very strongly with wavelength; note the λ^{-4} dependence of σ_R in equation (4.2). For a vertical column of atmosphere 40% or so is lost in the near ultraviolet while less than 1% is lost in the near infrared (problem 4.1). On average (taking into account different solar elevations and different wavelengths) about 13% of the solar radiation incident on the atmosphere is Rayleigh scattered (problem 4.2). Approximately half of this reaches the earth's surface as *diffuse* radiation the other half being returned to space.

Solar radiation is also scattered by particulate aerosol arising from a wide variety of sources including volcanoes, chimneys, dust from the earth's surface and sea spray. Such aerosol is present in the atmosphere in very variable quantities (problem 4.3).

4.2   Absorption of solar radiation by ozone

Reference to fig. 2.1 and fig. A8.1 will indicate that, in addition to attenuation through the scattering processes discussed in § 4.1 solar radiation is modified by absorption by atmospheric constituents – mainly by oxygen and ozone in the ultraviolet and by water vapour and carbon dioxide in the infrared. First we consider absorption of solar radiation by ozone in the stratosphere which is the major energy input to that region of the atmosphere.

Ozone is formed in the stratosphere (10–50 km) and mesosphere (50–80 km) by photochemical processes (§ 5.5), the peak of ozone concentration occurring at about 25 km altitude. Ozone absorbs ultraviolet solar radiation in the Hartley band (200–300 nm); at levels below 70 km virtually all the energy absorbed goes into kinetic energy of the molecules, i.e. into increasing the atmospheric temperature.

The absorption spectrum of ozone in the Hartley band is mainly a continuous spectrum, that is the absorption coefficient $k_{\tilde{v}}$ is a smoothly varying function of the wavenumber† $\tilde{v}$ and is also independent of pressure, so that, ignoring the complications of scattering which must be taken into account in an accurate calculation (cf. § 12.5), the incident solar flux $F_{S\tilde{v}}(z)$ at height z is given by

$$F_{S\tilde{v}}(z) = F_{S\tilde{v}}(\infty) \exp\left(-\int_z^\infty k_{\tilde{v}} \rho_z \sec\theta \, dz\right) \qquad (4.3)$$

where ρ_z is the density of ozone at height z, and θ is the zenith angle of the sun.

Note that in order to take account of variations with wavenumber, the quantity $F_{S\tilde{v}}(z)$ is the solar flux per unit wavenumber interval at wavenumber $\tilde{v}$.

To obtain the heating rate (cf. equation (2.4)) at level z an integration over frequency is required so that

† Throughout this chapter, frequencies will be expressed in wavenumbers denoted by $\tilde{v}$. The wavenumber is the reciprocal of the wavelength or the frequency in Hz divided by the velocity of light.

$$c_p \rho \frac{dT}{dt} = \cos\theta \int_{\text{band}} \frac{dF_{S\tilde{v}}}{dz} \, d\tilde{v} \tag{4.4}$$

Given values of $k_{\tilde{v}}$ across the spectrum and ρ_z at various heights, numerical integration of (4.3) and (4.4) can be carried out (see appendices 8 and 9 for information about solar flux and ozone absorption). For a typical ozone distribution at mid latitudes, the peak of heating is at ~ 50 km altitude; when integrated over a day, it amounts to a rate of change of temperature of ~ 8 K day^{-1} (problem 4.6).

4.3   Absorption by single lines

In the infrared part of the spectrum the absorption coefficient in molecular bands varies rapidly with frequency. Molecules possess discrete energy levels associated with different vibrational and rotational states; any given vibration–rotation band contains many thousands of individual lines (cf. fig. 4.1). The structure of the bands is often complex and is further complicated because each line in a band is broadened by collision broadening or Doppler broadening – broadening mechanisms which vary with temperature and pressure and which are, therefore, variable throughout the atmosphere.

In order to compute the average transmission of a path of atmosphere for a given spectral region, we need to sum over many individual lines.

The simplest model of a molecular band which may be employed is of a large number of lines which are independent (i.e. non-overlapping), broadening by collision processes only. For the carbon dioxide bands and water vapour bands of importance in the atmosphere this is a good approximation for levels in the stratosphere between ~ 20 km and 60 km altitude. Above these levels Doppler broadening becomes important (problem 4.7) and at lower levels lines overlap very substantially.

Simple theory leads to an absorption coefficient $k_{\tilde{v}}$ at wavenumber $\tilde{v}$ due to a collision broadened line centred at $\tilde{v}_0$ given by (see, for instance, Houghton & Smith, 1966; Eisberg, 1961)

$$k_{\tilde{v}} = \frac{s\gamma}{\pi\{(\tilde{v} - \tilde{v}_0)^2 + \gamma^2\}} \tag{4.5}$$

where $\quad s = \int_0^{\infty} k_{\tilde{v}} \, d\tilde{v} \tag{4.6}$

is the line strength and $\gamma = (2\pi t c)^{-1}$ the half-width (in wavenumbers). t is the mean time between collisions in the absorbing gas, a quantity which varies over several orders of magnitude in the atmosphere because it is inversely

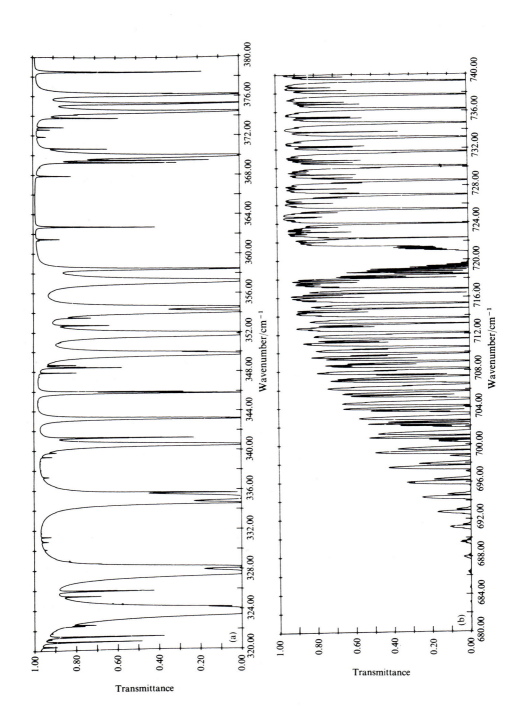

proportional to the pressure p. Ignoring the temperature dependence of γ which by comparison is small, we can write

$$\gamma = \gamma_0 \, p/p_0 \tag{4.7}$$

where γ_0 is its value at standard pressure p_0. For many gases, at STP, $\gamma_0 \simeq 0.1 \, \text{cm}^{-1}$, so that at 50 km altitude where the pressure is 0.1 kPa, $\gamma \simeq 10^{-4} \, \text{cm}^{-1}$.

Consider what appearance such a line would have when viewed with a high resolution laboratory spectrometer through a path of length l containing absorbing gas at density ρ. The transmission $\tau_{\tilde{\nu}}$ of such a path will be

$$\tau_{\tilde{\nu}} = \exp\left(-k_{\tilde{\nu}}\rho l\right) \tag{4.8}$$

and the *equivalent width* or *integrated absorptance* W of the line given by fig. (4.2)

$$W = \int_{-\infty}^{\infty} d\tilde{\nu}(1 - \tau_{\tilde{\nu}})$$
$$= \int_{-\infty}^{\infty} d\tilde{\nu}\{1 - \exp\left(-k_{\tilde{\nu}}\rho l\right)\} \tag{4.9}$$

Two approximations for W are very useful: (1) the *weak* approximation where the exponential in (4.9) may be approximated by the first term of its expansion, in which case on substituting for $k_{\tilde{\nu}}$ from (4.5) (problem 4.8)

$$W = s\rho l \tag{4.10}$$

and (2) the *strong* approximation for which there is no transmission of any consequence near the line centre (the line centre is then said to be black). The actual value of $k_{\tilde{\nu}}$ near the line centre is, therefore, not important providing it is large enough for γ^2 to be omitted from the denominator in (4.5). The integration in (4.9) may now be carried out (problem 4.9) and

$$W = 2(s\gamma\rho l)^{1/2} \tag{4.11}$$

Plotting the equivalent width W against the length of absorbing path l results in the *curve of growth* (fig. 4.2).

Taking in a substantial proportion of an absorption band of width $\Delta\tilde{\nu}$

Fig. 4.1. Illustrating the fine structure of molecular absorption bands. Atmospheric transmission for a 10 km horizontal path at 12 km altitude as computed by McClatchey & Selby (1972) for (a) the 320–380 cm^{-1} (31–26 μm) region where the lines are due to water vapour, (b) the 680–740 cm^{-1} (15–13 μm) region where the lines are mainly due to carbon dioxide.

Only a very small part of the spectrum is covered in this diagram. Compare with fig. 2.1 in which the variation of atmospheric transmission with frequency is on a very much coarser scale.

in which a number of non-overlapping lines are present, the average transmission $\bar{\tau}$ over the spectral interval will be

$$\bar{\tau} = 1 - \frac{\sum_i W_i}{\Delta \tilde{\nu}} \tag{4.12}$$

where W_i is the equivalent width of the ith line.

4.4 Transmission of an atmospheric path

We apply the results of the previous section to a vertical path in the atmosphere between levels of pressure p_1 and p_2 ($p_1 > p_2$). When the weak approximation applies this is straightforward as the equivalent width is then

Fig. 4.2. (a) Curve of growth of a typical spectral line with $s = 10^4 \, \text{cm}^{-1} \, (\text{g cm}^{-2})^{-1}$ and $\gamma_0 = 0.06 \, \text{cm}^{-1}$ showing the linear and square root regions of growth.

(b) Actual shapes of the spectral line for different values of ρl corresponding to the values shown by the arrows in (a).

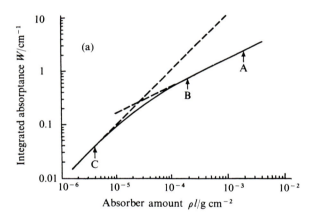

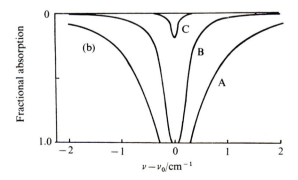

only a function of the amount of absorber in the path, i.e.

$$W = s \int_{\text{path}} c\rho \, dz \qquad (4.13)$$

where c is the fractional concentration (by mass) of absorber and ρ is the total density. Since $\rho \, dz = -dp/g$ and assuming a constant value of c along the path

$$W = sc(p_1 - p_2)/g \qquad (4.14)$$

It is not so clear how to proceed when the absorption is stronger, as the pressure (and hence γ) varies along the path. In this case application is made of the *Curtis–Godson approximation* which states that a mean pressure $\bar{p}$ may be defined for a path

$$\bar{p} = \frac{\int pc\rho \, dz}{\int c\rho \, dz} \qquad (4.15)$$

i.e. in deriving the mean pressure, the pressure along the path is weighted by the amount of absorber. This is an exact approximation when the strong approximation applies (problem 4.13); it is, of course, also exact under the weak approximation when the equivalent width is independent of pressure.

For uniform composition, $\bar{p} = \frac{1}{2}(p_1 + p_2)$ so that substituting from (4.7) and (4.15) in (4.11) we have in the strong approximation, for a single spectral line

$$W = 2\left\{ \frac{s\gamma_0 c}{2gp_0} (p_1^2 - p_2^2) \right\}^{1/2} \qquad (4.16)$$

in which case the average transmission over a spectral interval is, from (4.12) and (4.16)

$$\bar{\tau} = 1 - \frac{1}{\Delta\tilde{\nu}} \left\{ \frac{2c}{gp_0} (p_1^2 - p_2^2) \right\}^{1/2} \sum_i (s_i\gamma_{0i})^{1/2} \qquad (4.17)$$

where $\sum_i (s_i\gamma_{0i})^{1/2}$ is the sum of the square roots of the product of line strength and line width at STP within the spectral interval.

Provided the transmission is high the expression in (4.17) can be employed to work out the transmission for any given path or the proportion of solar radiation reaching different atmospheric levels (problem 4.15). A simple extension to the theory to cope with regions where lines overlap significantly will be presented in § 4.9.

4.5   The integral equation of transfer

So far in this chapter we have dealt with the scattering of solar radiation and with its transmission in absorbing regions in the ultraviolet and the infrared parts of the spectrum. We now need to consider for the infrared part of the spectrum both emission and absorption. We follow on from the discussion of radiation transfer in chapter 2 where a very simplified atmospheric model was considered. There the assumptions were made of an atmosphere in radiative equilibrium and of an absorption coefficient independent of frequency. Here, we shall set up the problem more generally showing how, for any given atmosphere, integrations over height and over frequency can be carried out. We shall then consider a particular case, but still confining the discussion to absorption and emission processes only, i.e. to a non-scattering atmosphere.

To evaluate the effect of radiative transfer in the atmosphere we need to be able to find the intensity of radiation at any given level in an atmosphere with any structure and composition. For the plane parallel atmosphere considered in §2.2 an expression for the intensity is obtained by integrating Schwarzschild's equation (2.3). This may readily be done using the integrating factor $\exp(-\chi)$ (problem 4.16). The integral equation may also be derived directly in a way which illustrates the physical processes involved.

Note that because it is necessary to take into account variations of absorption coefficient with frequency (or wavenumber) we first consider the transfer of radiation of intensity $I_{\tilde{v}}$ at wavenumber $\tilde{v}$, the units of $I_{\tilde{v}}$ being Watts per unit area per unit solid angle per unit wavenumber interval.

Consider a thick slab (fig. 4.3) between levels z_0 and z_1, with radiant intensity $I_{\tilde{v}0}$ incident vertically upwards at z_0. To calculate $I_{\tilde{v}1}$ the intensity

Fig. 4.3

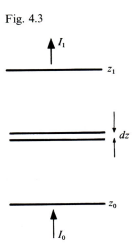

leaving the top of the slab, consider the elemental slab dz at level z ($z_0 < z < z_1$) and temperature T emitting radiation in the upward direction of intensity $k_{\tilde{v}}\rho\,dz\,B_{\tilde{v}}(z)$ (cf. § 2.2) where $k_{\tilde{v}}$ is the absorption coefficient at wavenumber $\tilde{v}$ and

$$B_{\tilde{v}} = 2\tilde{v}^2 \frac{hc\tilde{v}}{\exp(hc\tilde{v}/kT) - 1} \tag{4.18}$$

is the Planck function at wavenumber $\tilde{v}$ (cf. appendix 7), where the definition of the quantities is as in § 2.2. This emitted radiation will be attenuated before reaching z_1, the transmission (§ 2.2) being

$$\tau_{\tilde{v}}(z, z_1) = \exp\left(-\int_z^{z_1} k_{\tilde{v}}\rho\,dz'\right) \tag{4.19}$$

The contribution to $I_{\tilde{v}1}$ from the slab dz will be

$$dI_{\tilde{v}1} = k_{\tilde{v}}\rho\,dz\,B_{\tilde{v}}(z)\exp\left(-\int_z^{z_1} k_{\tilde{v}}\rho\,dz'\right)$$
$$= B_{\tilde{v}}(z)\,d\tau_{\tilde{v}}(z, z_1) \tag{4.20}$$

as may easily be verified by differentiating (4.19). We have, therefore,

$$I_{\tilde{v}1} = I_{\tilde{v}0}(z_0, z_1) + \int_{\tau_{\tilde{v}}(z_0, z_1)}^1 B_{\tilde{v}}(z)\,d\tau_{\tilde{v}}(z, z_1) \tag{4.21}$$

the first term being the contribution from the incident intensity at z_0.

Equation (4.21) is the integral equation of transfer and is important not only in radiation transfer within the atmosphere but also in the interpretation in terms of atmospheric structure of measurements of radiation leaving the top of the atmosphere (§ 12.4).

4.6 Integration over frequency

To obtain the total intensity of radiation at any given atmospheric level, it is necessary to integrate over all frequencies where significant radiation is present. Integrating (4.21) over frequency leads to

$$I_1 = \int_0^\infty I_{\tilde{v}0}\tau_{\tilde{v}}(z_0, z_1)\,d\tilde{v} + \int_0^\infty \int_{\tau_{\tilde{v}}(z_0, z_1)}^1 B_{\tilde{v}}(z)\,d\tau_{\tilde{v}}(z, z_1)\,d\tilde{v} \tag{4.22}$$

Of the quantities on the right hand side of (4.22), $B_{\tilde{v}}$ is a relatively slowly varying function of frequency whereas $\tau_{\tilde{v}}$ varies very rapidly with frequency (fig. 4.1). Given adequate information about the intensities, positions, widths and shapes of the spectral lines involved in any given spectral region, it is possible in principle to perform the integration over frequency in (4.22). However, to cope in practice with the integration it is necessary to find simplified methods of removing much of the complication associated with molecular band structure.

One such simplified method employing the non-overlapping line approximation was discussed in § 4.3; further methods will be presented in § 4.9 and problem 4.19.

4.7   Heating rate due to radiative processes

The main reason why radiative transfer calculations are required is to deduce the contribution to the atmosphere's energy budget from radiative processes. The local heating rate is expressed in (2.4) in terms of the divergence of the upward and downward radiation fluxes.

For nearly all atmospheric situations it is adequate to employ the simple approximation of § 2.2 for integration over angle, so that if in (4.22) τ is replaced by τ^* (i.e. pathlengths are multiplied by $\frac{5}{3}$) and B is replaced by πB, an equation for the upward flux $F^{\uparrow}$ at level z_1 can be written

$$F^{\uparrow} = \int_0^{\infty} F_{\tilde{v}0} \tau_{\tilde{v}}^*(z_0, z_1)\, d\tilde{v} + \int_0^{\infty} \int_{\tau_{\tilde{v}}^*(z_0, z_1)}^{1} \pi B_{\tilde{v}}(z)\, d\tau_{\tilde{v}}^*(z, z_1)\, d\tilde{v} \qquad (4.23)$$

An entirely similar expression applies for the downward flux $F^{\downarrow}$.

4.8   Cooling by carbon dioxide emission from upper stratosphere and lower mesosphere

We now apply the non-overlapping strong collision-broadened line approximation to the v_2 band of carbon dioxide at 15 μm wavelength in the 20–60 km region. In this part of the atmosphere the temperature distribution is mainly determined by a balance between radiative cooling in the infrared from this carbon dioxide band and heating by absorption of solar radiation by ozone (§ 4.2).

To compute the radiation loss and hence the cooling rate we shall employ the *cooling to space* approximation, in which exchange of radiation between different atmospheric layers is neglected in comparison with the loss of radiation direct to space – a good approximation for this particular atmospheric region. Equating the loss of energy to space, the term $dF^{\uparrow}/dz$ from (2.4), to the cooling rate we have

$$\frac{dT}{dt} \rho c_p = - \int_{\Delta\tilde{v}} \pi B_{\tilde{v}}(T) \frac{d\tau_{\tilde{v}}^*(z, \infty)}{dz}\, d\tilde{v} \qquad (4.24)$$

where the integration over frequency is as in (4.23) and covers the region $\Delta\tilde{v}$ of the 15 μm carbon dioxide band. Over this spectral interval $B_{\tilde{v}}(T)$ can be considered constant with frequency. From (4.17)

$$\int_{\Delta\tilde{v}} \tau_{\tilde{v}}^*(z, \infty)\, d\tilde{v} = \Delta\tilde{v} - p \left(\frac{10c}{3g p_0} \right)^{1/2} \sum_i (s_i \gamma_{0i})^{1/2} \qquad (4.25)$$

The factor $\frac{5}{3}$ is included because we require the slab transmission appropriate to flux calculations. The strong approximation has also been applied to all the lines in the band; the result of problem 4.20 shows that this is a valid simplification. Substituting in (4.24) and remembering that $\rho\, dz = -dp/g$ we have

$$\frac{dT}{dt} = -\frac{g\pi B_{\tilde{\nu}}(T)}{c_p}\left(\frac{10c}{3gp_0}\right)^{1/2}\sum_i (s_i\gamma_{0i})^{1/2} \tag{4.26}$$

Substituting values $c_p = 1005\ \mathrm{J\ kg}^{-1}$, $c = 4.7 \times 10^{-4}$, $\sum_i(s_i\gamma_{0i})^{1/2} = 1600\ \mathrm{cm}^{-1}$ $(\mathrm{g\ cm}^{-2})^{-1/2}$ (appendix 10), we have

$$\frac{dT}{dt} = 2.0\pi B_{\tilde{\nu}}\ \mathrm{K\ s}^{-1} \tag{4.27}$$

where $\pi B_{\tilde{\nu}}$ is in $\mathrm{W\ cm}^{-2}(\mathrm{cm}^{-1})^{-1}$. Note the important result that the factor multiplying $B_{\tilde{\nu}}$ in the expression for the cooling rate is independent of height.

A first approximation to the temperature near 50 km altitude may be obtained by assuming that radiative equilibrium applies and that the heating due to the absorption of solar radiation by ozone (§ 4.2) is balanced by the cooling due to CO_2 emission. Inserting the ozone heating rate from problem 4.6 in equation (4.27), the equilibrium value of $\pi B_{\tilde{\nu}}(T)$ at 50 km is found to be $4.6 \times 10^{-5}\ \mathrm{W\ cm}^{-2}\ (\mathrm{cm}^{-1})^{-1}$ equivalent to a temperature T of $\sim 290\ \mathrm{K}$, approximately as found at that level. A diurnal temperature variation of $\sim 4\ \mathrm{K}$ is present at 50 km because heating only occurs during the day but cooling goes on all the time.

4.9 Band models

The simplest model of a band is that of non-overlapping lines described in § 4.3. Such a model, however, is only applicable when the lines are well separated. For the atmosphere it can only be applied for the path lengths and pressures above altitudes of about 30 km. A variety of band models have been invented which allow for overlap, in particular the regular model first described by Elsasser, in which an absorption band is simulated by an array of equally spaced lines of the same shape and strength, and the random model due to Goody in which the lines are considered to be spaced randomly and to have a distribution of line strengths following some statistical law. This latter model leads to a particularly simple and useful expression for the average transmission $\bar{\tau}$ of a spectral interval of width $\Delta\tilde{\nu}$ containing a large number of lines

$$\bar{\tau} = \exp\left(-\sum W_i/\Delta\tilde{\nu}\right) \tag{4.28}$$

where $\sum W_i$ is the sum of the equivalent widths of all the lines in the interval considered as independent lines. In problem 4.19 the theory of the random band model is developed further. From this theory and from the spectral data provided in appendix 10, the transmission of any spectral interval in the bands of water vapour, carbon dioxide and ozone in the region of the thermal infrared can be calculated.

With the use of band models, the frequency integration of (4.22) may be carried out for rather few spectral intervals, each major absorption band being divided into just one or two spectral regions. Such simplified radiation calculations are required for incorporation into numerical models (§ 11.7).

4.10   Continuum absorption

In the wavelength region 8–13 μm in the infrared, apart from the ozone band at 9.6 μm, the atmosphere is reasonably transparent (fig. 2.1). Some absorption by water vapour lines is still present, however, and of more importance, by water vapour dimers, i.e. associations of water vapour molecules in pairs. This latter absorption is a continuum with absorption coefficient proportional to the partial pressure of water vapour; associations are more likely to occur when the partial pressure is high. For this reason the continuum absorption is particularly important in humid atmospheres; for a wet tropical atmosphere, transmission of a vertical path through the atmosphere may only be $\sim 50\%$ in this region (problem 4.17).

4.11   Global radiation budget

In this chapter, equations have been written down which, given detailed information regarding the spectra of the molecules, enable calculations of the radiation fluxes and the radiation heating rate to be made at any level in a clear atmosphere. No account has been taken of clouds and their radiative properties. Clouds are, in fact, probably the dominant influence in the radiative budget of the lower atmosphere but adequately taking them into account raises many problems, which will be considered briefly in § 6.4.

Of fundamental importance to our understanding of the atmosphere's energy budget is adequate knowledge of the components of the radiation budget at the top of the atmosphere. Fig. 4.4 shows what happens on average to the incident solar radiation. The outgoing long-wave radiation originates from near the surface or from cloud top in the window region and from high in the troposphere in the strong carbon dioxide and water vapour bands (cf. fig. 12.3). Fig. 4.5 shows the results of satellite measurements of the two main

Fig. 4.4. The interactions of incident solar radiation with the earth and its atmosphere.

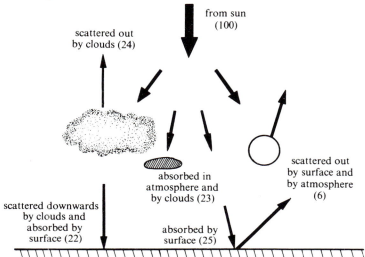

Fig. 4.5. Average components of the earth's radiation budget as deduced from satellite observations, 1962–66, by Vonder Haar & Suomi (1971).

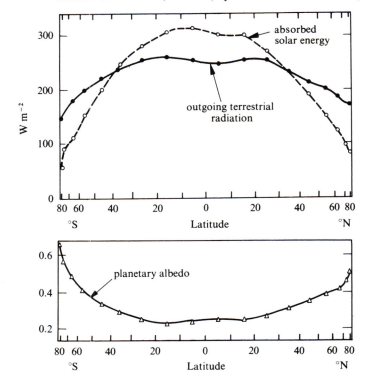

components of the radiation budget averaged over longitude and over a complete year. The excess of solar energy absorbed over that emitted in equatorial regions must be transported through atmospheric motions or ocean currents to the polar regions where it is required to supply the deficit between energy emitted and received. How atmospheric motions achieve the transfer will be discussed in chapter 10.

Problems

4.1 From the expression (4.2) for the scattering coefficient for Rayleigh scattering at wavelength λ, and using the data in appendix 1, calculate the extinction by Rayleigh scattering for a vertical column of atmosphere at wavelengths of 0.3, 0.6 and 1 μm.

4.2 Continue the calculations of problem 4.1 for a few other wavelengths and for a number of solar elevations. Hence estimate, for a clear atmosphere, with Rayleigh scattering only, the proportion of solar radiation scattered out by the atmosphere from the whole hemisphere illuminated by the sun.

4.3 The Angström turbidity factor of an atmosphere containing aerosol is defined as the optical thickness τ_A of a vertical column due to aerosol scattering. Assuming that the scattering cross-section of the particles is twice their geometric cross-section, estimate what density of aerosol of diameter 1 μm would be needed in the lowest kilometre of the atmosphere to produce a turbidity factor of unity.

4.4 It is commonly assumed that the Angström turbidity factor varies with wavelength λ as $\lambda^{-1.3}$. For an atmosphere with an optical depth due to aerosol of 0.3 at 0.6 μm, what is the value of the turbidity coefficient at 0.3 μm and 1 μm respectively. From the results of problem 4.1, calculate the total extinction due to Rayleigh and aerosol scattering for such an atmosphere at the three wavelengths.

4.5 Explain why scattered sunlight observed at 90° to the direction of the sun shows very strong polarization. Why is the polarization only about 90%, not 100%?

4.6 Assuming the sun overhead and a uniform temperature atmosphere, for a distribution of the number density n_3 of ozone molecules as a function of pressure p given by $n_3 = n_0 p^{3/2}$ (quite a good approximation to the upper part of the ozone layer), show that the heating rate h by absorption of solar radiation is of the form

$$\frac{h}{h_m} = \left(\frac{p}{p_m}\right)^{1/2} \exp\left\{-\frac{1}{3}\left[\left(\frac{p}{p_m}\right)^{3/2} - 1\right]\right\} \tag{4.29}$$

where p_m is the level of maximum heating h_m. In your calculation assume a single absorption coefficient independent of wavelength within the ozone band.

Given that the level of maximum heating is at 0.1 kPa pressure, the solar flux over the ozone absorbing region is 0.7×10^{-3} W cm^{-2}, calculate h_m.

4.7 Because of the ellipticity of the earth's orbit the solar radiation incident on the earth is about 7% greater in December than in June. Assuming constant ozone, from the variation with temperature of the Planck function at 15 μm, estimate the temperature difference between the temperature of the stratopause (the level of maximum temperature near 50 km) in the summer over the north pole and over the south pole. Compare your result with the observed temperature difference of about 2.5 K (Barnett (1974)).

4.8 The shape of a spectral line due to Doppler broadening is given by

$$k_{\tilde{v}} = \frac{s}{\gamma_D \pi^{1/2}} \exp\left\{-\left(\frac{\tilde{v} - \tilde{v}_0}{\gamma_D}\right)^2\right\} \tag{4.30}$$

where $\quad \gamma_D = \frac{\tilde{v}_0}{c}\left(\frac{2RT}{M_r}\right)^{1/2} \tag{4.31}$

(M_r is molecular weight).

Calculate the pressure in the atmosphere at which the half-width due to collision broadening is equal to that due to Doppler broadening for (1) a carbon dioxide line at 667 cm^{-1}, (2) a water vapour line at 1600 cm^{-1}, (3) a water vapour line at 100 cm^{-1}. Assume all lines have collision broadened half-widths of 0.1 cm^{-1} at standard pressure.

For the two shapes compare the ratio of the absorption coefficient at the line centre to its value several half-widths away. Notice that the absorption coefficient in the wings of collision broadened lines is relatively much greater than that for the wings of Doppler broadened lines. For this reason, even for conditions where the Doppler half-width is greater than the collision broadened half-width the transfer of radiation in the collision broadened wings of the lines remains important.

4.9 Integrate (4.9) under the weak approximation to obtain (4.10).

4.10 Integrate (4.9) under the strong approximation to obtain (4.11).

4.11 Consider a path of fixed length l in which the pressure of an absorbing gas can be varied. Show that under conditions of collision broadening the absorption at the line centre is independent of pressure.

4.12 Show that for a single collision broadened line the integral of (4.9) may be written as

$$W = \int_{-\infty}^{\infty} dx \left\{ 1 - \exp\left(\frac{-2u\gamma^2}{x^2 + \gamma^2}\right) \right\}$$

where $x = \tilde{v} - \tilde{v}_0$, $u = s\rho l/2\pi\gamma$

The value of this integral may be expressed in terms of the modified Bessel functions $I_n(u) = -iJ_n(iu)$ giving

$$W = 2\pi\gamma lu \exp(-u)\{J_0(u) + I_1(u)\}$$
$$= 2\pi\gamma L(u) \tag{4.32}$$

The function $L(u)$ is known as the *Ladenberg and Reiche function*.

4.13 Substitute the strong approximation to $k_{\tilde{v}}$ from (4.5) (i.e. omit the γ^2 in the denominator) into (4.19) for the transmission along an atmospheric path and show that the Curtis–Godson approximation (4.15) is exact in the case of the strong approximation.

4.14 Two useful approximations to the equivalent width W of a collision broadened line are

$$W = s\rho l \left(1 + \frac{s\rho l}{4\gamma}\right)^{-1/2}$$

and

$$W = s\rho l \left\{ 1 + \left(\frac{s\rho l}{4\gamma}\right)^{5/4} \right\}^{-2/5}$$

Show that these expressions have the correct 'weak' and 'strong' limits ((4.10) and (4.11)). The first deviates by less than 8% from the correct equivalent width for all values of the parameters. The second due to Goldman (1968) has a maximum error of about 1%.

4.15 From equations (4.17) and (4.28) and the data of Appendices 2, 3 and 10 estimate the average transmission of a vertical path from the top of the atmosphere to the 20 kPa level for the spectral interval 525–550 cm^{-1}. Also for latitude 60° N estimate the transmission of a vertical path of atmosphere down to the surface for the spectral interval 775–800 cm^{-1}.

4.16 Integrate (2.3) to derive (4.21) using the integrating factor $\exp(-\chi)$.

4.17 Absorption in the atmospheric window between $8\,\mu m$ and $13\,\mu m$ is mostly due to the water vapour dimer; the absorption coefficient is of the form $k_2 e$ where e is the water vapour pressure (in kPa), $k_2 \simeq 10^{-1}$ $(g\,cm^{-2})^{-1}\,kPa^{-1}$. If the water vapour pressure near the surface is $1\,kPa$, calculate (1) the transmission of a horizontal path $1\,km$ long near the surface, (2) the transmission of a vertical path of atmosphere assuming that the distribution of water vapour pressure is proportional to (pressure in atmospheres)^4. Also estimate for a layer near the surface the cooling rate in K day^{-1} by radiation from water vapour in this spectral region.

4.18 Integrate (4.21) by parts to give

$$I_{\tilde{v}1} = \{I_{\tilde{v}0} - B_{\tilde{v}}(z_0)\}\tau_{\tilde{v}}(z_0, z_1) + B_{\tilde{v}}(z_1) + \int_{z_1}^{z_0} \tau_{\tilde{v}}(z, z_1)\frac{dB_{\tilde{v}}(z)}{dz}\,dz$$

Write this equation in terms of flux and using (2.4) show that when $I_{\tilde{v}0} = B_{\tilde{v}}(z_0)$ an expression for the atmospheric heating rate at level z_1 is

$$\rho c_p \frac{dT}{dt} = \pi \int_0^{\infty} \int_0^{\infty} \frac{d\tau_{\tilde{v}}(z, z_1)}{dz_1} \cdot \frac{dB_{\tilde{v}}(z)}{dz}\,dz\,d\tilde{v}$$

4.19 Suppose in a random array of spectral lines the number $N(s)\,ds$ of lines having a strength between s and $s+ds$ is

$$N(s)\,ds = \frac{N_0}{s}\exp\left(-\frac{s}{\sigma}\right) \tag{4.33}$$

This distribution is known as the Malkmus model (Malkmus, 1967) and fits well to line distributions found in practice. Show that the sum of the equivalent widths $\sum W_i$ of these lines (assuming no overlap) after passing through a path of length l with absorber density ρ is

$$\sum W_i = \int_0^{\infty} \frac{N_0}{s}\exp\left(-\frac{s}{\sigma}\right)ds \int_{-\infty}^{\infty} \left\{1 - \exp\left(-k_{\tilde{v}}\rho l\right)\right\}dz \tag{4.34}$$

If a line shape is described by

$$k_{\tilde{v}} = sf(\tilde{v}) \tag{4.35}$$

carry out the integration over s to give

$$\sum W_i = N_0 \int_{-\infty}^{\infty} \ln\{1 + \sigma\rho l f(\tilde{v})\}\,d\tilde{v} \tag{4.36}$$

If $k_{\tilde{v}}$ is given by the collision broadened expression (4.5) show that

$$\sum W_i = 2\pi\gamma N_0\left\{\left(1 + \frac{\sigma\rho l}{\pi\gamma}\right)^{1/2} - 1\right\} \tag{4.37}$$

Compare the values of $\sum_i W_i$ given by (4.37) in the limits of $\sigma\rho l/\pi\gamma \ll 1$ and $\gg 1$ with what would be expected from the weak and strong approximations respectively (4.10) and (4.11)

$$\sum_i W_i \,(\text{weak}) = \sum_i s_i \rho l \tag{4.38}$$

$$\sum_i W_i \,(\text{strong}) = 2 \sum_i (s_i \gamma_i \rho l)^{1/2} \tag{4.39}$$

where the summation in each case is over all the lines denoted by strength s_i and half-width γ_i.

The best way of fitting spectral line data to the random model is not by attempting to match the model distribution of line strengths with the actual distribution but rather by matching the weak and strong limits of (4.37) with (4.38) and (4.39) respectively. By doing this, find expressions for $\pi N_0 \gamma$ and $\sigma/\pi\gamma$ in (4.37) in terms of $\sum s_i$ and $\sum (s_i \gamma_i)^{1/2}$. Hence show that (4.37) may be written

$$\sum W_i = \frac{2\{\sum (s_i \gamma_i)^{1/2}\}^2}{\sum s_i} \left[\left\{ 1 + \frac{\rho l (\sum s_i)^2}{[\sum (s_i \gamma_i)^{1/2}]^2} \right\}^{1/2} - 1 \right] \tag{4.40}$$

This is a particularly useful form of $\sum W_i$ which, for any given spectral interval $\Delta\tilde{v}$ can be substituted in (4.28) to calculate the transmission under any given circumstance when collision broadening applies. Spectral line data for many bands of importance suitable for substituting in (4.40) are listed in appendix 10.

4.20 For calculations of the transmission of a vertical path of atmosphere between a level where the pressure is p and the top of the atmosphere, show, by applying the data of appendix 10 for the 15 μm carbon dioxide band to (4.40) that, providing collision broadening applies, the strong approximation is a very good assumption.

5

The middle and upper atmospheres

5.1 Temperature structure

Fig. 5.1 shows a typical cross-section of the temperature structure of the earth's atmosphere from the surface up to ~ 100 km. The temperature falls with increasing height in the lowest 10 or 15 km roughly at the adiabatic lapse rate, as discussed in chapter 2. Above the tropopause the main features of the temperature structure are determined by radiative processes. The high temperature at ~ 50 km (known as the *stratopause*) is due to absorption of solar radiation by ozone (§ 4.2) – balanced mainly by emission of infrared radiation by carbon dioxide. Above the temperature peak at 50 km is the *mesosphere* (middle sphere) where the temperature falls with height, although not so steeply as in the troposphere. The upper boundary of the mesosphere is the temperature minimum around 80 km altitude known as the *mesopause* where there is rather little absorption of solar radiation. Above the mesopause, solar ultraviolet radiation is strongly absorbed, particularly by molecular and atomic oxygen, and the temperature rises rapidly to between 500 K and 2000 K in the region known as the *thermosphere*.

The variation of temperature with latitude should be noted from fig. 5.1. At the earth's surface and at the level of the stratopause the equator is warmer than the polar regions, as simple considerations would predict. The tropopause and the mesopause are, however, substantially colder over the equator than over the poles. This is particularly surprising for the winter polar regions where there is no solar radiation. Radiative considerations cannot provide an explanation of these reversed temperature gradients. Their existence demonstrates that atmospheric motions not only act as heat engines and transfer energy from source regions to sink regions but can also act as refrigerators transferring energy in the other direction.

In the following sections of this chapter we shall consider various physical processes which occur in the atmosphere above 30 km. Only 1% of the atmospheric mass lies above 30 km and only 1 part in 10^5 above ~ 80 km. It must not be expected, therefore, that events in these upper regions have a very immediate or large effect on the structure or motion of the troposphere. However, because of the close links which exist between motions in various parts of the atmosphere, study of upper atmospheric motion can have an important bearing on our understanding of motion lower down. Further, what happens at the top of the atmosphere is important in the study of the long-term evolution of the atmosphere. Consideration, therefore, of upper atmosphere phenomena, in addition to being of interest in its own right, plays an important part in the understanding of the whole.

5.2   Diffusive separation

The amount of mixing in the lower atmosphere ensures that no significant diffusive separation occurs between the heavy atmospheric constituents (e.g. argon) and the light ones (e.g. hydrogen). For practical purposes, therefore, the main constituents of the lower atmosphere are uniformly mixed apart from those which are involved in phase changes (e.g. water vapour) or chemical changes (e.g. ozone).

Although turbulent mixing dominates over molecular diffusion in the lower atmosphere, because the molecular diffusion coefficient is proportional to the mean free path and hence inversely proportional to density, in the very high atmosphere molecular diffusion plays the dominant role both for the transfer of the molecules themselves and for the transfer of heat. A typical value of molecular diffusion coefficient for an atmospheric gas is $\sim 10^{-4}$ m^2 s^{-1} at STP and therefore ~ 100 m^2 s^{-1} at a pressure of 10^{-6} atm (corresponding to ~ 96 km altitude). This is of the same order as a typical eddy diffusion coefficient (cf. §9.2) appropriate to the lower thermosphere. The lower boundary of the region above which molecular diffusion dominates is ~ 120 km altitude and is known as the *turbopause* (sometimes known as the *homopause* : the region above, where molecular diffusion is dominant, is sometimes known as the *homosphere*). The turbopause can often be observed fairly clearly from the luminous trails of rockets, as a boundary below which the trail is violently disturbed by turbulence and above which it is relatively smooth.

We first consider the composition of an atmosphere in a state of diffusive equilibrium. In such equilibrium the hydrostatic equation (1.4) is applied to each constituent separately. This follows from Dalton's law of partial pressures, together with the condition that there be no net transport of any constituent across a horizontal surface. For the case of an isothermal atmosphere

$$\ln \{ p(i)/p_0(i) \} = -z/H(i) \tag{5.1}$$

where the index i refers to the ith constituent. If $T = 1000$ K, the order of temperature found in the thermosphere, the values of the scale height H for argon ($M_r = 40$), nitrogen ($M_r = 28$), atomic oxygen ($M_r = 16$) and atomic hydrogen ($M_r = 1$) are respectively 21 km, 30 km, 54 km and 850 km. From (5.1), the ratio of the number densities n of two different constituents at an altitude z may be compared to their number densities n_0 at altitude z_0 at which diffusive separation begins. For example, for argon and nitrogen,

$$\frac{n(A)n_0(N_2)}{n_0(A)n(N_2)} = \exp\left[-(z - z_0)\left\{ \frac{1}{H(A)} - \frac{1}{H(N_2)} \right\} \right] \tag{5.2}$$

The results of such calculations are shown in fig. 5.2. Whereas nitrogen is the most abundant constituent in the lower atmosphere, this role is taken over by atomic oxygen at about 170 km and by helium at about 500 km, these altitudes being quoted for mean atmospheric conditions; they vary very considerably with atmospheric temperature (problem 5.1).

5.3   The escape of hydrogen

As density decreases in the atmosphere with increasing altitude, eventually a level is reached where collisions are so rare that a molecule moving upwards will turn about under the influence of gravity and return again to the denser layers without making a collision. Under such conditions the faster molecules escape from the atmosphere altogether. The region is called the *exosphere*. The *critical level* z_c for such escape may be defined such that a proportion e^{-1} of a group of very fast particles moving vertically upwards at z_c will experience no collisions as they pass completely out of the atmosphere. For molecules of diameter d, the probability of no collision being made in a vertical path dz is $\exp[-\pi d^2 n(z)\, dz]$, where $n(z)$ is the particle density at height z. Substituting for $n(z)$ from the hydrostatic equation (1.4) the condition for the critical level z_c is therefore

$$\int_{z_c}^{\infty} \pi d^2 n(z_c) \exp\left[\frac{-(z - z_c)}{H} \right] dz = 1 \tag{5.3}$$

i.e. $n(z_c) = (\pi d^2 H)^{-1}$ (5.4)

The departure from equilibrium is sufficiently small for the scale height H to be considered approximately constant through the region. Since $[n(z)\pi d^2]^{-1}$ is the mean free path in the horizontal direction at level z, the critical level may

Fig. 5.1. Mean latitudinal cross-sections of temperature (K), surface to ∼ 100 km, for northern hemisphere for (a) summer and winter, (b) spring and autumn. Data for the diagrams tabulated in appendix 5 where the sources of data are listed.

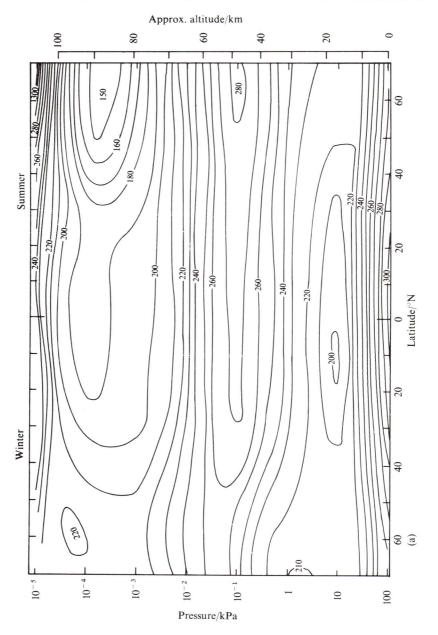

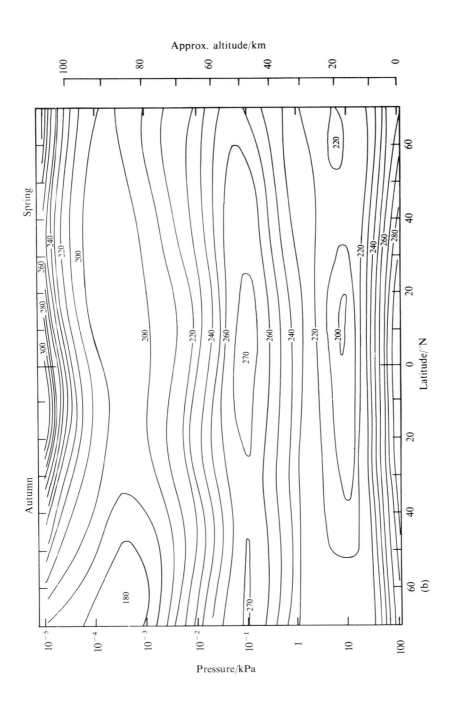

Fig. 5.1. (c) mean January and July zonal winds (from COSPAR, 1972). The relation between the temperature and wind fields is described in § 7.6.

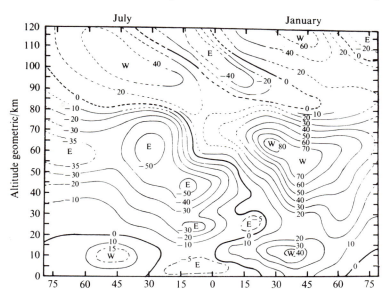

(c)   Fig. 5.1 (c).

Fig. 5.2. Total number density and number densities of various constituents for mean atmospheres. (From COSPAR, 1972)

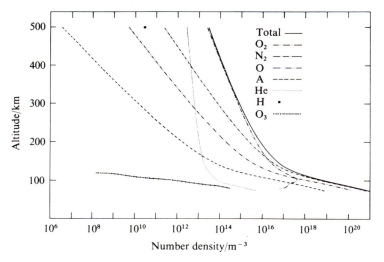

also be defined as that at which the mean free path in the horizontal direction is equal to the scale height H. The critical level occurs at an altitude where the most abundant constituent is atomic oxygen (fig. 5.2). For a temperature of 1000 K (approximate value for mean solar activity (fig. 5.3)) its scale height H is 53 km. With $d \simeq 2 \times 10^{-10}$ m, $n(z_c) \simeq 10^{14}$ m^{-3} corresponding to a height z_c of between 400 and 500 km (fig. 5.2). Considerable variations, of course, occur with solar activity, time of day and season.

We now consider the rate of escape of hydrogen atoms. These originate mainly from molecules of water vapour and methane which are dissociated in the stratosphere and mesosphere by the action of ultraviolet radiation from the sun. For escape upwards from the earth's gravitational field a particle of mass m and velocity V must possess a kinetic energy $\frac{1}{2}mV^2$ greater than its gravitational potential energy mGM/a (G is the gravitational constant and M and a respectively the mass and radius of the earth), i.e.

$$\frac{1}{2}mV^2 > \frac{mGM}{a}$$

or $\qquad V^2 > 2ga, \quad \text{since } g = \frac{GM}{a^2}$ \hfill (5.5)

Fig. 5.3. Mean temperature of the upper atmosphere during periods of (a) very low, (b) mean, (c) very high, solar activity (from COSPAR, 1972). An indication of variations with time of day is given in fig. 5.4.

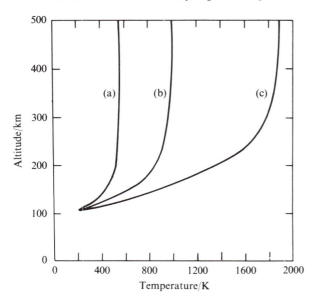

or $V > V_c$

where $V_c = 11 \text{ km s}^{-1}$

This is over twice the mean molecular velocity of atomic hydrogen at a temperature of 1000 K, and very much larger than the mean velocities of other molecules.

Since the rate of escape is small the velocity distribution at the escape level will be approximately Boltzmann. The number dn_H of hydrogen atoms per unit volume having velocities between V and $V + dV$ is (see, for instance, Jeans, 1940)

$$dn_H = \frac{n_H(z_c) V^2 \, dV \exp(-\alpha V^2)}{\int_0^\infty V^2 \, dV \exp(-\alpha V^2)} \tag{5.6}$$

where $\alpha = m/2kT$ and m is the mass of a hydrogen atom. Of these the number moving in an upward direction across unit area per second is $\frac{1}{4} V \, dn_H$ (see, for instance, Jeans, 1940), so that the number $\dot{N}$ of atoms escaping per unit time from unit area is

$$\dot{N} = \frac{1}{4} n_H(z_c) \left(\frac{\alpha^3}{\pi}\right)^{1/2} \int_{V_c}^\infty V^3 \, dV \exp(-\alpha V^2) \, dV \tag{5.7}$$

where a further factor $\frac{1}{4}$ has been included to allow for the fact that the number density at the critical level was computed for travel vertically upwards whereas (5.7) includes molecules travelling in any upward direction. The integral in (5.7) may be evaluated by parts to give

$$\dot{N} = \frac{1}{8} n_H(z_c) \left(\frac{\alpha}{\pi}\right)^{1/2} \left(V_c^2 + \frac{1}{\alpha}\right) \exp(-\alpha V_c^2) \tag{5.8}$$

$$= \beta n_H(z_c), \text{ say} \tag{5.9}$$

where β has the dimensions of a velocity.

For hydrogen and $T = 1000$ K, $\beta \simeq 1.6 \text{ m s}^{-1}$.

The flux $\dot{N}$ of hydrogen atoms has, of course, to be maintained right through the atmosphere up to the escape level. Above the turbopause, it is maintained predominantly by molecular diffusion, the net flux of atoms at any level due to diffusion being (omitting the rather small thermal diffusion term)

$$D\left\{\frac{dn_H}{dz} - \left(\frac{dn_H}{dz}\right)_e\right\} \tag{5.10}$$

where D is the diffusion coefficient and $(dn_H/dz)_e$ the vertical gradient of n_H under equilibrium conditions which from the hydrostatic equation is $-n_H/H_H$, H_H being the scale height for hydrogen atoms. Carrying through the calculation (problem 5.5) shows that to provide the required upward flux of hydrogen atoms, the ratio of hydrogen atom concentration at 120 km to that at the escape level must be substantially greater than it would be if no escape occurred.

5.4   The energy balance of the thermosphere

Above the turbopause at ~ 120 km altitude, we have already seen in § 5.2 that, because of the low density and the positive temperature gradient with altitude, molecular diffusion is more important than eddy diffusion. The same is true for heat transfer. The dominant processes controlling the temperature of the thermosphere are the absorption of solar radiation in the extreme ultraviolet (wavelengths below 91 nm) by atomic oxygen, balanced by molecular conduction of heat. Of lesser importance under normal conditions are radiative transfer by far infrared transitions within the ground state of atomic oxygen and transfer of energy from particles which make up the *solar wind*. Under disturbed conditions the latter can become the dominant source of energy.

Ignoring, for the moment, variations through the day, consider equilibrium conditions for the atmospheric layer above level z. Since the conduction of heat across the top of the atmosphere is zero the solar radiation absorbed within a layer averaged over a day must be equal to the conduction of heat out of the bottom of the layer, i.e.

$$\varepsilon\{\bar{I}(\infty) - \bar{I}(z)\} = \lambda(z)\,\frac{dT(z)}{dz} \qquad (5.11)$$

where $\bar{I}(z)$ is the average incident solar radiation, $\lambda(z)$ the coefficient of thermal conductivity and $T(z)$ the temperature, all at level z. The factor ε allows for the fact that not all the energy absorbed will be transformed into heat; some will be re-radiated out of the region altogether and some of the ionized or dissociated particles will diffuse out of the region to recombine elsewhere. According to simple kinetic theory, λ is proportional to the mean molecular velocity $\bar{c}$ which varies as $T^{1/2}$ but is independent of the pressure.

Knowing the distribution of atomic oxygen (§ 5.2) and its ultraviolet properties and also making some assumption about the value of ε, (5.11) may be solved numerically (problem 5.7). Fig. 5.3 shows the result of such calculations for different solar activities.

The problem of thermospheric heat balance may also be written in time-dependent form so that the diurnal variation may be estimated. Fig. 5.4 shows the variation of thermospheric temperature through the day as derived from such a calculation.

5.5   Photochemical processes

We have already seen the importance of ozone in determining the temperature structure of the stratosphere and mesosphere. The most important photochemical processes in the atmosphere, therefore, are those which lead to the formation and destruction of ozone. The simple 'classical' theory of ozone due to Chapman (1930) considered reactions involving oxygen only, namely:

$$O_2 + h\nu \rightarrow O + O \qquad\qquad (J_2) \qquad\qquad (5.12)$$

$$O + O + M \rightarrow O_2 + M \qquad\qquad (k_1) \qquad\qquad (5.13)$$

$$O + O_2 + M \rightarrow O_3 + M \qquad\qquad (k_2) \qquad\qquad (5.14)$$

$$O + O_3 \rightarrow 2O_2 \qquad\qquad (k_3) \qquad\qquad (5.15)$$

$$O_3 + h\nu \rightarrow O_2 + O \qquad\qquad (J_3) \qquad\qquad (5.16)$$

Equations (5.12) and (5.16) describe photodissociation in the presence of solar radiation, the first occurring for wavelengths less than 246 nm and the second for wavelengths less than 1140 nm but most strongly below 310 nm. J_2 and J_3 are dissociation rates per molecule per second, which will, of course, vary with altitude. k_1, k_2 and k_3 are reaction rates defined such that for reactions between two species present in number densities n_1 and n_2 respectively per unit volume, the number of reactions per second will be the product $n_1 n_2$ multiplied by the reaction rate. Reactions (5.13) and (5.14) are three-body collisions, the third body M being required to satisfy energy and momentum conservation simultaneously. For these reactions to obtain the number of

Fig. 5.4. The variation over the equator at the equinox of exospheric temperature with local time for a case when the minimum temperature is 1000 K. (From COSPAR, 1972)

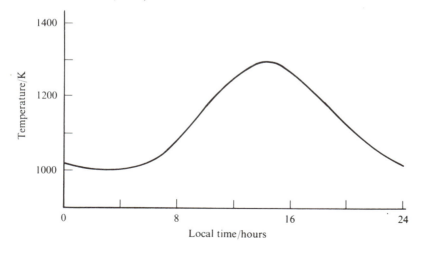

reactions per second, the appropriate reaction rate is multiplied by the product of the number densities of the three species involved.

In the stratosphere reaction (5.13) is slow and may be neglected. Reactions (5.14) and (5.16), in which 'odd' oxygen particles (i.e. O and O_3) are converted into each other, are much faster than reactions (5.12) and (5.15) in which 'odd' oxygen particles are created or destroyed. Because of this, in the stratosphere, equilibrium between the O and O_3 concentration is governed by (from the law of Mass Action)

$$J_3 n_3 = k_2 n_2 n_1 n_M \tag{5.17}$$

where n_1, n_2, n_3 and n_M are respectively the concentrations of O, O_2, O_3 and all molecules.

The balance between creation and destruction of odd oxygen particles is described by

$$2J_2 n_2 = 2k_3 n_1 n_3 \tag{5.18}$$

From (5.17) and (5.18) the equilibrium ozone concentration is

$$n_3 = n_2 \left(\frac{J_2 k_2 n_M}{J_3 k_3} \right)^{1/2} \tag{5.19}$$

The rates J_2 and J_3 depend differently on altitude (fig. 5.5), J_2 dropping off with decreasing altitude more rapidly than J_3, mainly because of the overlap of ozone and oxygen absorption in the 200 nm region. A peak of ozone

Fig. 5.5. Values of dissociation rates J_2 and J_3 averaged over a day for typical mid-latitude conditions. (From Crutzen, 1971)

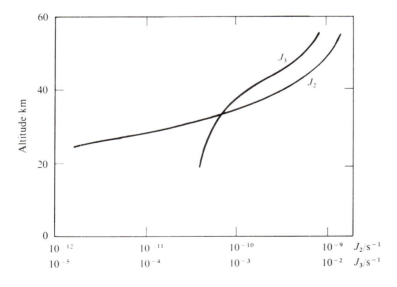

concentration, therefore, occurs (fig. 5.6). Agreement between the computed ozone profile and that observed is not very good, because (1) other reactions involving minor constituents such as NO, NO_2 or other particles such as H, OH, HO_2 themselves formed photochemically are involved in the formation and destruction of O and O_3 and because (2) the lifetime of ozone molecules at levels below 30 km is long (problem 5.9) so that ozone is redistributed by atmospheric motions. We shall look briefly at these effects in turn.

Regarding the influence of minor constituents, processes removing 'odd' oxygen from the stratosphere can be visualized in terms of catalytic cycles

Fig. 5.6. (a) Height distribution of ozone at 45° latitude, summer (A) as observed and (B) as calculated from 'classical' pure oxygen photochemical theory (after Dütsch, 1968). (b) Total ozone in a vertical column (expressed as the depth of the column of gas if isolated from the rest of the atmosphere and at STP in units of 10^{-5} m, sometimes known as Dobson units) plotted as a function of latitude and season (after Dütsch, 1971). Note the maxima in spring at high latitudes in both hemispheres which result from transport by atmospheric motions.

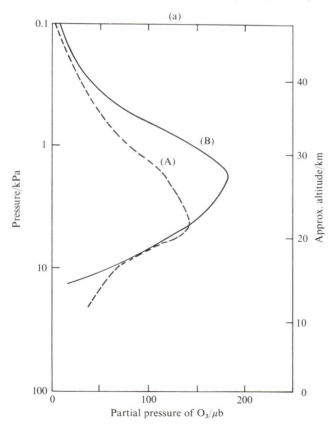

in which the conversion of a species X to XO and back again destroys two 'odd' oxygen particles; i.e.

$$X + O_3 \rightarrow XO + O_2 \qquad (5.20)$$

followed by

$$XO + O \rightarrow X + O_2 \qquad (5.21)$$

leads to the net result of

$$O + O_3 \rightarrow 2O_2 \qquad (5.22)$$

Important species X involved in such catalytic cycles in the stratosphere are H, NO and Cl (problem 5.17).

Because the distribution of these constituents, especially NO and Cl, may be influenced by man's activities, and because of concern about the destruction of part of the atmospheric ozone layer which might ensue through these catalytic reactions, a lot of attention has been given to them in recent years. The chemistry, however, is far from simple; the radical species we have mentioned react rapidly with each other as well as with ozone, and under some

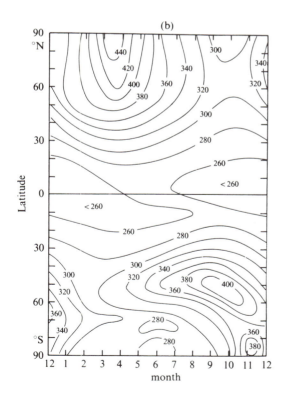

circumstances more ozone is created than destroyed by adding appropriate species.

The second large influence on the ozone distribution mentioned above is that of atmospheric motion. Reference to fig. 5.6 will illustrate the effect. Although on photochemical theory it would be expected that the maximum ozone would occur in tropical regions where there is maximum solar radiation and a minimum in polar regions, in fact the reverse is the case. To account for this Brewer (1949) proposed what has become known as the *Dobson–Brewer* circulation in which air enters the stratosphere through upward motion in the tropics (cf. § 10.9). Ozone-rich air is then transported polewards and downwards so that a large concentration of ozone occurs at high latitudes in the lower stratosphere where its photochemical lifetime is very long. Observations of the ozone mixing-ratio provide an important tracer of these motions. During the winter a great deal of planetary wave activity is observed in the stratosphere (§ 8.6) the effect of which on transport processes needs to be taken into account for a complete picture of stratospheric behaviour to be realized (§ 10.9).

Some photochemical reactions give rise to excited molecular species. The resulting radiation is known as *airglow*. A particularly important reaction is

$$H + O_3 \rightarrow OH^* + O_2 \qquad\qquad (5.23)$$

which occurs near the mesopause (~ 90 km) and leads to vibrationally excited OH molecules which radiate in the near infrared. The amount of emission is large and makes a significant contribution to the energy budget of the mesopause region (problem 5.10).

5.6   Breakdown of thermodynamic equilibrium

In the discussion of radiative transfer in chapters 2 and 4 the assumption of local thermodynamic equilibrium (LTE) enabled Kirchhoff's law to be applied so that the Planck black-body function could be employed as the source function in the equation of transfer (2.3). At high levels in the atmosphere LTE is no longer a good assumption, and the molecular processes involved need to be considered in more detail.

Consider a system (fig. 5.7) of the ground state and first excited state of a particular vibrational mode. At atmospheric temperatures very few molecules will be in states higher than the first and we shall neglect these higher states in this simplified treatment. Molecules can be excited into the upper level either by absorbing radiation of the appropriate frequency or through the effect of collisions. Molecules can lose their vibrational energy and transfer back to the ground state by emitting radiation either spontaneously or by stimulated emission or again through collisions. If the volume of gas under consideration is completely enclosed so that there is no net gain or loss by radiation a Boltzmann balance between the rate of excitation and de-excitation will be provided through the radiation and collision processes considered separately, and the ratio of the population of the upper level to that of the lower level will be the Boltzmann factor $\exp(-h\nu/kT)$. If now part of the enclosure is removed so that some net radiaton is gained or lost from the system a Boltzmann balance cannot be maintained through absorption and emission processes. However, it is still possible that the rates of excitation and de-excitation by collision will be sufficiently rapid that they dominate over the radiation processes; the population of the upper level will then be affected only a little by the radiative gain or loss. This is the situation of local thermodynamic equilibrium. We may expect it no longer to apply when (1) there is a significant net radiative gain or loss from the excited level in question and when (2) the rates of excitation or de-excitation by collisions are comparable with the rates of excitation or de-excitation by absorption or emission.

To set up the problem, the first treatment of which was by Milne (1930), we need to express parameters describing the radiation field such as the absorption coefficient and the source function in terms of the detailed molecular rate constants of fig. 5.7.

First considering emission, the radiation power emitted by the element of volume dv (fig. 5.8) containing $n\,dv$ molecules, between frequencies ν and $\nu + d\nu$, and in solid angle $d\omega$ is $k_\nu nJ_\nu\,ds\,dA\,d\nu\,d\omega$ where k_ν is the absorption cross-section per molecule – an expression which serves to define the source function J_ν. Assuming that J_ν is isotropic and, over the spectral region occupied by the absorption band, does not vary with ν, since $\int_{\Delta\nu} k_\nu\,d\nu = S$ (the total band

Fig. 5.7. A two level system with Einstein coefficients: A_{21} for spontaneous emission, B_{12} for absorption, B_{21} for stimulated emission, a_{21} and b_{12} are probabilities per unit time of relaxation or excitation by collisional processes.

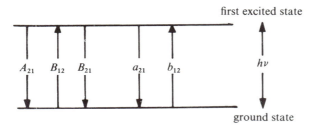

strength of the vibrational transition, i.e. integrated over all rotational structure within the whole band of width Δv), the total power emitted by the element is $4\pi SnJ_v \, dv$.

Turning this expression for total power emitted into a number of photons per unit time enables it to be related to the Einstein coefficient A_{21} (fig. 5.7), i.e.

$$4\pi SnJ_v(hv)^{-1} = A_{21}n_2 \tag{5.24}$$

Note that in (5.24) we have neglected stimulated emission compared with spontaneous emission as for the bands of interest under atmospheric conditions spontaneous emission dominates.

In a similar way absorption of radiation in the element of volume dv can be considered. If I_v is the incident intensity within the solid angle $d\omega$ and the spectral interval dv, the power absorbed will be $k_v I_v n \, dv \, dv \, d\omega$ which after integration gives, for the total power absorbed,

$$4\pi Sn \, dv \, \bar{I}_v$$

where $\quad \bar{I}_v = \dfrac{1}{4\pi S} \displaystyle\int_{\Delta v} \int_{4\pi} k_v I_v \, d\omega \, dv \tag{5.25}$

i.e. the intensity averaged with respect to ω and v. Now the rate of absorption of photons per unit volume is also equal to $B_{12}n_1\rho_v$ where ρ_v is the energy density of radiation at frequency v within the volume dv. Since $\rho_v = 4\pi\bar{I}_v c^{-1}$, we therefore have

$$4\pi Sn\bar{I}_v(hv)^{-1} = B_{12}n_1 4\pi c^{-1}\bar{I}_v \tag{5.26}$$

Referring again to fig. 5.7 equilibrium between the population of the levels requires

$$n_1(B_{12}4\pi\bar{I}_v c^{-1} + b_{12}) = n_2(A_{21} + a_{21}) \tag{5.27}$$

Substituting in (5.27) for n_1 and n_2 from (5.24) and (5.26) results in an expression

Fig. 5.8

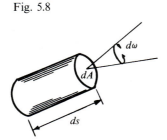

for the source function

$$J_v = \frac{\bar{I}_v + (cb_{12}A_{21}/4\pi a_{21}B_{12})\phi}{1+\phi} \tag{5.28}$$

where $\phi = a_{21}/A_{21}$

In the limit where local thermodynamic equilibrium applies collisional activation and deactivation dominates so $\phi \to \infty$ and $J_v \to B_v$, the Planck function. The quantity in brackets in (5.28) must therefore be B_v and

$$J_v = \frac{\bar{I}_v + \phi B_v}{1+\phi} \tag{5.29}$$

Note that as $\phi \to 0$, $J_v \to \bar{I}_v$ and we have isotropic scattering.

If ψ is written for the heating rate, or the rate of increase in energy per unit volume due to radiation processes

$$\psi = -\int_{\Delta v} \int_{4\pi} \frac{dI_v}{ds} \, d\omega \, dv \tag{5.30}$$

Substituting for dI/ds from the equation of transfer (2.3) (with J_v written for B_v):

$$\psi = \int_{\Delta v} \int_{4\pi} k_v n (I_v - J_v) \, d\omega \, dv$$

$$= 4\pi S n (\bar{I}_v - J_v) \tag{5.31}$$

from (5.25). Substituting for $\bar{I}_v$ from (5.29) we have

$$J_v - B_v = \frac{\psi}{4\pi S n \phi} \tag{5.32}$$

Equation (5.32) expresses the difference between the source function and the black-body function in terms of the heating rate and the ratio ϕ of the probability of relaxation from the upper state through collisional processes a_{21} to that through radiation processes. As we should expect from the discussion at the beginning of this section, $J_v \simeq B_v$, i.e. LTE exists when the heating rate is small or when collisional processes dominate, i.e. when ϕ is large.

The most important application in the atmosphere of the theory of non-LTE radiative transfer is to emission by the $15 \, \mu m$ vibration–rotation band of carbon dioxide, which for the LTE case was considered in § 4.8. With the same approximation as was employed there, that cooling to space is the dominant term, for the non-LTE case (4.24) applies with B_v replaced by J_v i.e.

$$-\psi = \int_{\Delta v} \pi J_v \frac{d\tau_v^*(z, \infty)}{dz} \, dz \tag{5.33}$$

with $\tau_v^*(z, \infty) = \exp\left(-\int_z^\infty \frac{5}{3} k_v n \, dz'\right)$ (5.34)

Over the spectral interval in which the band is contained, J_v may be considered constant and

$$-\psi = \tfrac{5}{3}\pi J_v n S \overline{\tau_v^*}$$ (5.35)

where the quantity

$$\overline{\tau_v^*} = \frac{1}{S} \int_{\Delta v} k_v \tau_v^*(z, \infty) \, dv$$ (5.36)

is the probability of a photon emitted by a carbon dioxide molecule at level z getting out to space. Substituting for J_v from (5.32) in (5.35) we have

$$-\psi = \frac{\tfrac{5}{3}\pi S n \overline{\tau_v^*} \, B_v}{1 + 5\overline{\tau_v^*}/12\phi}$$ (5.37)

Under conditions of LTE, the second term in the denominator of (5.37) is zero. The degree to which the cooling rate, therefore, differs from its LTE value depends not only on ϕ but also on $\overline{\tau_v^*}$ the transparency of the atmosphere to space.

For the 15 μm carbon dioxide band, and collisions with air molecules, the collisional relaxation time $a_{21}^{-1} \simeq 3 \times 10^{-5}$ s at standard pressure and 210 K (an average temperature for the mesosphere). Since a_{21} is mainly dependent on the collision frequency it is approximately proportional to pressure. The radiative life time $A_{21}^{-1} = 0.74$ s. We therefore have $\phi \simeq 2.4 \times 10^4 p$ (where p is pressure in atm) and $\phi = 1$ at a pressure of 4 Pa or a height of 73 km. Because, at this level $\overline{\tau_v^*}$ is small (problem 5.12), the LTE approximation applies to somewhat higher altitudes, in fact to about the 80 km level. Above 80 km the second term in the denominator becomes more and more significant. For instance at the 10^{-6} atm level (~ 96 km), the rate of cooling is only 0.05 of what it would be under LTE (problem 5.14 and fig. 5.9).

In the above discussion, the cooling to space approximation has been used. We conclude this section by writing down the heating rate equations in a particularly useful form for numerical computation in which the radiative transfer between all layers is included correctly.

Consider the atmosphere divided into a number of discrete layers. With the plane parallel approximation of § 4.7 and for a frequency range sufficiently wide to include much or all of a vibration–rotation band, but sufficiently narrow that B_v or J_v is constant within the range, (2.4) and (4.23) can be written in finite difference form

$$\psi_k = \sum_j C_{kj} J_{vj} + \psi_{Sk} \tag{5.38}$$

where ψ_k is the heating rate at level k, J_{vj} is the source function at level j, C_{kj} is an element of a matrix known as the *Curtis matrix* which depends on atmospheric transmission and the summation is over all atmospheric layers. Heating, due to solar absorption ψ_{Sk} has been included to make the treatment as general as possible. In the same notation (5.32) is

$$J_k = B_k + E_k \psi_k \tag{5.39}$$

where $E_k = (4\pi S n \phi)^{-1}$ appropriate to level k. In matrix notation, (5.38) and

Fig. 5.9. Heating rates for the 15 μm band of carbon dioxide for a mean atmosphere, with collisional relaxation times at standard pressure (a) 2×10^{-6} s, (b) 1×10^{-5} s, (c) 3×10^{-5} s. (After Williams, 1971)

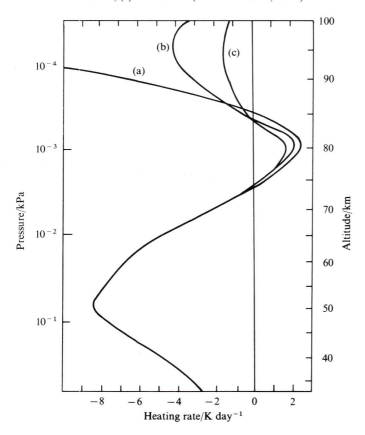

(5.39) are

$$\Psi = CJ + \Psi_S$$

$$J = B + E\Psi \tag{5.40}$$

which may be solved to give

$$\Psi = (I - CE)^{-1}(CB + \Psi_S) \tag{5.41}$$

I being the unit matrix.

In (5.41) **C** describes atmospheric transmission between different levels which for a uniformly mixed gas such as carbon dioxide may be computed once and for all for a variety of standard atmospheric conditions (the dependence of transmission on temperature is small). The vector **B** describes a particular atmospheric temperature profile. The form of (5.41), therefore, is particularly convenient for numerical computation of radiative heating rates for different atmospheric conditions.

Problems

5.1 Assuming different uniform temperatures above the turbopause at 120 km, estimate for different levels of solar activity the level above which the most abundant constituent is helium.

5.2 Show that a body of cross-sectional area A travelling at velocity V through a gaseous medium of density ρ experiences a drag force equal to $\frac{1}{2}\rho A c_D V^2$ where c_D is the drag coefficient. Hence estimate for standard atmospheric conditions the change in orbital period per orbit for a satellite of mass 100 kg with $A = 1 \, \text{m}^2$, $c_D = 2$, in a circular orbit at 600 km altitude. Observation of the details of satellite orbits is an important method of measuring density in the thermosphere.

5.3 Calculate the value of β (5.9) for helium atoms for $T = 1000$ K. Estimate the temperature which would be necessary for the value of β for helium to be 1.6 m s^{-1}, i.e. equal to that for hydrogen at $T = 1000$ K.

5.4 Show that, from simple kinetic theory, the diffusion coefficient D of hydrogen atoms present in small proportion in air at temperature T and total number density n is proportional to $T^{1/2}n^{-1}$.

Hence show that if a constant scale height H is assumed for the atmosphere as a whole above the turbopause where the total number density is n_0 then the density of hydrogen atoms is given by the differential equation

$$\frac{dn_H}{dz} + \frac{n_H}{H_H} = -\frac{\dot{N}n_0}{CT^{1/2}} \exp\left\{\frac{-(z - z_0)}{H}\right\} \tag{5.42}$$

where C is a constant.

5.5 Integrate (5.42). From your solution, given that n_H (120 km) $= 2 \times 10^{11}$ m^{-3}, $C = 2 \times 10^{20}$ m^{-1} s^{-1} K$^{-1/2}$ and finding suitable values for the other quantities from tables or diagrams, find the value of n_H at the escape level and hence the rate of escape of hydrogen atoms.

5.6 It has been suggested that oxygen present in the earth's atmosphere has resulted from the dissociation of water molecules in the high atmosphere followed by the escape of hydrogen. Assuming constant conditions calculate the loss of hydrogen during the earth's history ($\sim 4.5 \times 10^9$ years) and test the plausibility of this hypothesis. (This possible source is now believed to be small compared with oxygen produced through photosynthesis.)

5.7 Solve (5.11) numerically in the following way (you will need to write a simple computer programme). Assume a pure oxygen atmosphere above a lower boundary at $z = 120$ km, $T = 300$ K and calculate from the hydrostatic equation the number density of atomic oxygen as a function of altitude for an isothermal atmosphere. Assume an absorption cross-section for atomic oxygen of 1.2×10^{-21} m^2, a value of the product εI_∞ of 1.1×10^{-3} W m^{-2}, a value of $\lambda = AT^{1/2}$ where $A = 3.6 \times 10^{-3}$ J m^{-1} s^{-1} K$^{-3/2}$ and solve (5.11) for layers beginning with the lower boundary. From the resulting values of T, recalculate a new density profile and continue the iteration until a satisfactory solution has been obtained.

5.8 For the temperature profile of appendix 6 calculate the total potential energy (thermal and gravitational) above 120 km as given by (3.19). Compare this with the energy absorbed during an eight hour period (assuming the sun overhead). Hence make a crude estimate of the amplitude of diurnal variation in the exosphere. Compare your estimate with fig. 5.4.

5.9 Associated with the equilibrium of ozone as described in § 5.5, there are two very different time constants, the first being associated with changes in the relative amount of O and O_3, the combined number densities of O and O_3 being kept constant, and the second being associated with changes in the sum of the amounts of O and O_3, the ratio between them during the change being given by (5.17). Write down differential equations for changes under these two approximations and hence deduce expressions for the two time constants. Calculate values for them at 40 km, 30 km and 20 km, given that $k_2 = 10^{-33}$ cm^6 s^{-1}, $k_3 = 10^{-15}$ cm^3 s^{-1}. Take values of J_2, J_3 and n_3 from figs. 5.5 and 5.6.

Note that at the lower levels the second time constant is very long indeed so that the ozone mixing ratio becomes a very good 'tracer' of atmospheric motions.

5.10 Measurements of the OH airglow show that approximately 10^{12} photons s^{-1} of wavelength between 1 μm and 3 μm are emitted from a 1 cm^2 column of atmosphere at levels near the mesopause. Assuming that the emission is uniformly distributed over the region from 75 to 95 km altitude, estimate the cooling rate in K day^{-1} due to this emission.

5.11 If W is the equivalent width of an absorption band for the path between level z and space (absorber path length $u = |\int \frac{5}{3} c\rho \, dz|$), show that $\overline{\tau_v^*}$ in (5.36) is given by

$$\overline{\tau_v^*} = \frac{1}{S} \frac{dW}{du}$$

where S is the strength of the band.

Hence show that for a band of a gas present in constant mixing ratio consisting of non-overlapping collision broadened lines under either the 'weak' or 'strong' approximation the probability of a photon emitted from a level z getting out to space is independent of height.

5.12 Calculate the value of $\overline{\tau_v^*}$ for a photon from the 15 μm carbon dioxide band under the non-overlapping 'strong' collision broadened approximation given that $\sum (S_i \gamma_{0i})^{1/2} = 1600$ cm^{-1} (g cm^{-2})$^{-1/2}$ and $S = \sum S_i = 1.26 \times 10^5$ cm^{-1} (g cm^{-2})$^{-1}$ and that the mass mixing ratio of carbon dioxide $= 4.7 \times 10^{-4}$.

5.13 Consider cooling from the 15 μm carbon dioxide band at the level where the pressure is 10^{-6} atm (~ 96 km). Here $\phi \ll 1$ and $\overline{\tau_v^*} \approx 1$. Substitute for B_v and for A_{21} ($A_{21} = 8\pi v^2 g_1 S / c^2 g_2$ where g_1 and g_2 are the statistical weights of the lower and upper levels respectively) in (5.37). Hence show that

$$-\psi = nhv \left\{ \frac{g_2 a_{21}}{g_1 [\exp (hv/kT) - 1]} \right\}$$

Notice that this result is independent of the band strength S, and that the quantity multiplying nhv is the probability per unit time of a quantum of energy being acquired on collision. Put in values appropriate to the 10^{-6} atm level and find the cooling rate in K per day.

5.14 Compare the value found in problem 5.13 with that which would occur if thermodynamic equilibrium prevailed.

5.15 Check the assumption $\overline{\tau_v^*} = 1$ for the lines in the 15 μm carbon dioxide band considered in problem 5.13 by finding the transmission at the centre of a typical line of strength $s = 10^3$ cm^{-1} (g cm^{-2})$^{-1}$ between the level 10^{-6} atm and space. (The absorption coefficient at the centre of a Doppler broadened line is $s\pi^{-1/2}\gamma_D^{-1}$ – see problem 4.8 for an expression for γ_D).

5.16 Calculate for carbon dioxide at STP (molecular diameter 4×10^{-10} m) the number of collisions per molecule per second from simple kinetic theory considerations. If the relaxation time for de-excitation of the v_2 vibration by collision is 6×10^{-6} s at STP, what is the probability of de-excitation on one collision?

 This very small probability is because $h\nu$ for the v_2 vibration is considerably greater than the average kinetic energy of a molecule $(\sim \frac{3}{2}kT)$ at ordinary temperatures so that being able to transfer enough energy from translational motion to vibrational motion in any one collision is very unlikely.

5.17 Compare the rate of destruction of ozone at 40 km by the catalytic cycle of equations (5.20) and (5.21) with $X = Cl$ with destruction through equation (5.15). Take the concentration of ClO to be 10^8 cm^{-3}, the rate constant of 5.20 with $X = Cl$ to be 5×10^{-11} m^3 s^{-1} and other data as given in problem 5.9.

5.18 The Dobson–Brewer circulation mentioned in § 5.5 was also introduced to account for the very low water vapour content of the stratosphere. Supposing air always enters the stratosphere through the tropical tropopause estimate from the tables in appendices 2 and 5 what value of water vapour mixing ratio can be expected in the stratosphere.

6

Clouds

6.1 Cloud formation

It was demonstrated in § 3.2 that ascent of damp air can lead to condensation and hence cloud formation. Four main kinds of clouds can be distinguished according to the type of ascending motions which produce them: (1) layer clouds formed by widespread, regular ascent such as occurs near the boundary between air masses having different characteristics (frontal zones) or because of orography; (2) layer clouds formed by vertical mixing (cf. problem 3.9); (3) convective clouds; (4) clouds formed under stable conditions by limited vertical motion as air passes over hills or mountains. The two first processes lead to *stratus* clouds or, when at higher levels, *cirrus* clouds, convective processes produce *cumulus* clouds, and associated with mountains are typically *lenticular* or *wave* clouds. Examples of these different cloud types will be found in the pictures taken from orbiting satellites in figs. 3.4, 3.5, 3.6 and 7.9.

Suitable nuclei on which condensation can occur are reasonably abundant in the atmosphere. Nuclei on which the freezing of a droplet may commence are, however, less abundant; the tops of clouds extending above the freezing level commonly contain supercooled droplets at temperatures well below 0°C.

The presence of clouds influences the energetics of the atmosphere in two main ways: (1) by the part which clouds play in the atmospheric water cycle; latent heat is released on condensation and liquid water is removed from the atmosphere on precipitation; (2) by scattering, absorption and emission of solar and terrestrial radiation clouds influence very strongly the atmosphere's radiation budget. In this chapter these two main influences will be discussed in turn.

6.2 The growth of cloud particles

A cloud particle (droplet or ice crystal) has a typical diameter of

$\sim 10\,\mu\text{m}$ whereas a rain drop of sufficient size to precipitate is $\sim 1\,\text{mm}$ in diameter. Each precipitating particle, therefore, contains the same amount of liquid water as about one million cloud particles. Two possible processes are available for particle growth: (1) by diffusion of water vapour to the cloud particles and subsequent condensation; (2) by collision and coalescence of particles.

First consider growth by condensation of an isolated spherical particle of mass m, radius r, density ρ_L in an environment where at distance x from the particle the vapour density is ρ. The rate of particle growth is sufficiently slow for a steady state diffusion equation to be applied to it in which the mass of water vapour crossing any spherical surface in unit time is independent of x and equal to $\dot{m}$, the rate of increase of mass of the particle.

From Fick's law of diffusion we have, therefore,

$$\dot{m} = 4\pi x^2 D \frac{d\rho}{dx} \qquad (6.1)$$

where D is the diffusion coefficient of water vapour in air. Integrating (6.1) from the surface of the drop where the vapour density is ρ_r to a large distance away where it is ρ_∞ we have

$$4\pi D \int_{\rho_\infty}^{\rho_r} d\rho = \int_{\infty}^{r} \frac{\dot{m}}{x^2}\, dx \qquad (6.2)$$

As water vapour condenses on the particles, latent heat is released at a rate $L\dot{m}$ where L is the latent heat of condensation. This has to be conducted away. A temperature gradient dT/dx is therefore, set up, the equation describing the conduction being similar to (6.1), namely

$$L\dot{m} = -4\pi x^2 \lambda \frac{dT}{dx} \qquad (6.3)$$

where λ is the thermal conductivity of air. Equation (6.3) may be integrated to give

$$L\dot{m} = 4\pi \lambda r (T_r - T_\infty) \qquad (6.4)$$

where T_r and T_∞ are respectively the temperatures at the surface of the drop and at a large x.

From (6.2) and (6.4) the rate of droplet growth under different conditions can be evaluated (problems 6.1, 6.2 and 6.3). The process is particularly important at temperatures $\sim -10°\text{C}$ when a few ice crystals are present in a cloud which predominantly contains supercooled water drops (cf. fig. 3.6). The saturation vapour pressure over the water drops (see appendix 2) is about 10% more than that over the ice crystals which, therefore, grow at the expense of the water drops.

The other process of droplet growth is by collision and coalescence for which the rate of growth of a larger droplet of fall speed V in a population of smaller drops of fall speed v will be

$$\frac{dr}{dt} = \frac{\varepsilon w}{4\rho_L}(V-v) \qquad (6.5)$$

where w is the mass of liquid water in the form of smaller droplets per unit volume of the cloud and ε is the collision efficiency. Since small droplets which approach each other tend to follow the streamlines of the air, they are unlikely to collide unless they are both large enough and unless there is sufficient size differential between them. Cloud particles grow to $30\,\mu$m in diameter largely by condensation processes after which the collision efficiency becomes larger, subsequent growth continuing by coalescence (fig. 6.1).

6.3   The radiative properties of clouds

In chapter 2 when considering the simple radiative equilibrium model the assumption was made that solar radiation incident on the atmosphere is unmodified by the atmosphere and all absorbed at the surface; although the absorption of a small proportion of the solar radiation by ozone in the ultraviolet and by water vapour in the infrared were mentioned (cf. fig. 2.1). Solar radiation is also modified on its passage through the atmosphere by

Fig. 6.1. Rate of growth of cloud droplets for cloud with liquid water content $1\,\mathrm{g\,m}^{-3}$ (a) by coalescence assuming that distribution function of cloud droplet radii r given by $4r^2r_m^{-3}\exp(-2r/r_m)$ with $r_m = 2\,\mu$m, (b), (c), (d) by condensation assuming water vapour pressure is above saturation by (b) 4 Pa, (c) 1.2 Pa, (d) 0.25 Pa: after Shishkin (see Matveev, 1967, p. 513).

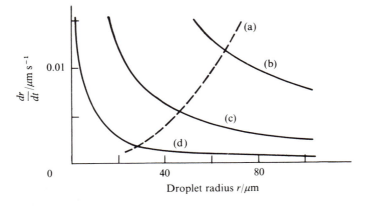

scattering processes (§ 4.1). Clouds are also of great importance; because on average they cover ~ 50% of the earth's surface their influence on both the incoming and outgoing radiation is large.

Considering first the modification of solar radiation by clouds (fig. 6.2), if F_S is the incident solar flux on the top of the cloud, AF_S the reflected upward flux at the top of the cloud (A being the *cloud albedo*) and τF_S the downward flux at the bottom of the cloud (τ *being its transmissivity*), the amount of radiant energy absorbed is $F_S(1 - A - \tau)$. For a typical stratus cloud and for radiation integrated over the whole spectrum, typical values are $A = 0.5$, $\tau = 0.3$. Calculations of how A and τ depend on cloud thickness and on the properties of the cloud particles are presented in § 6.4.

Because liquid water and ice absorb strongly throughout the infrared region, clouds are strong absorbers of terrestrial radiation. For considerations of the atmosphere's radiation budget it is satisfactory to assume that all clouds except cirrus absorb and emit terrestrial radiation as black bodies. Cirrus clouds are often thin and tenuous; although their albedo is small their transmissivity and emissivity may vary very considerably.

6.4   Radiative transfer in clouds

In chapters 2 and 4 radiative transfer theory was developed considering absorption and emission only. A simple extension of the theory enables scattering also to be taken into account.

Considering scattering particles distributed uniformly through a volume, a scattering coefficient σ can be defined in the same way as the absorption coefficient (2.1) so that the element of optical depth $d\chi = (k + \sigma)\rho \, dz$.

The absorption coefficient k takes into account absorption both by the particles and by the gas in between them. The coefficient σ accounts for scattering in all directions. A quantity called the *albedo for single scattering* ω is the ratio $\sigma/(k + \sigma)$.

Fig. 6.2

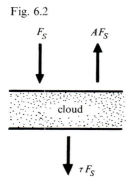

As in chapters 2 and 4 we are considering transfer in the vertical dimension only (integrating over other angular dependences) with fluxes $F^\uparrow$ and $F^\downarrow$. The transfer equations become

$$\frac{dF^\downarrow}{d\chi^*} + F^\downarrow = \pi B(1-\omega) + \omega\{fF^\downarrow + (1-f)F^\uparrow\} \tag{6.6}$$

$$-\frac{dF^\uparrow}{d\chi^*} + F^\uparrow = \pi B(1-\omega) + \omega\{(1-f)F^\downarrow + fF^\uparrow\} \tag{6.7}$$

The first term on the right hand side is the emission term depending on the black-body function B in exactly the same way as in (2.3). The second term is the contribution to radiation transfer from the scattering processes. The fraction f is the proportion of the scattered radiation which in one scattering event goes into a forward direction while $(1-f)$ is that which is scattered in a backward direction. For isotropic scattering $f=\frac{1}{2}$.

Equations (6.6) and (6.7) account properly for absorption, emission and for multiple scattering in cases where the restriction to one dimension is a satisfactory approximation. Their solution in the general case can be complicated. Here we shall treat only the case for solar radiation incident vertically downwards on the top of a cloud; the emission term (the first on the right hand side) in (6.6) and (6.7) will therefore be omitted.

It is instructive first to let $\omega=1$ when we have pure scattering. Equations (6.6) and (6.7) may then be written

$$\frac{dF^\downarrow}{d\chi^*} + (F^\downarrow - F^\uparrow)(1-f) = 0$$

$$\frac{dF^\uparrow}{d\chi^*} + (F^\downarrow - F^\uparrow)(1-f) = 0 \tag{6.8}$$

giving immediately

$$F^\downarrow - F^\uparrow = \text{constant}$$

and for a uniform cloud of optical thickness χ_0^* the boundary conditions are (fig. 6.2) $F^\downarrow (\chi^*=0) = F_S$, $F^\uparrow (\chi^*=\chi_0^*) = 0$ so that the cloud albedo A and transmissivity τ respectively become

$$A = \frac{\chi_0^*(1-f)}{1+\chi_0^*(1-f)} \tag{6.9}$$

$$\tau = \frac{1}{1+\chi_0^*(1-f)} \tag{6.10}$$

Note that $A+\tau=1$ as is required by conservation of energy.

When $\omega \neq 1$ (6.6) and (6.7) may be solved by differentiating again (6.6) and then substituting for $dF^\uparrow/d\chi^*$ from (6.7). The solution is

$$F^\downarrow = C \exp(-\alpha\chi^*) + D \exp[-\alpha(\chi_0^* - \chi^*)]$$

with a similar expression for $F^\uparrow$ where

$$\alpha^2 = (1-\omega)(1+\omega-2\omega f) \qquad\qquad (6.11)$$

and C and D are constants. Applying the boundary conditions leads to

$$A = \beta\, \frac{1 - \exp(-2\alpha\chi_0^*)}{1 - \beta^2 \exp(-2\alpha\chi_0^*)} \qquad\qquad (6.12)$$

$$\tau = \frac{(1-\beta^2)\exp(-\alpha\chi_0^*)}{1 - \beta^2 \exp(-2\alpha\chi_0^*)} \qquad\qquad (6.13)$$

where $\quad \beta = \dfrac{\alpha - 1 + \omega}{\alpha + 1 - \omega}$ $\qquad\qquad (6.14)$

Note that as χ_0^* becomes large

$$A \to \beta \quad \text{and} \quad \tau \to 0$$

The quantity β, therefore, is the albedo of a very thick cloud. For $\omega = 0.9997$, $f = 0.9$, values appropriate at a wavelength of $\sim 0.7\,\mu$m to a stratus cloud with drops $\sim 10\,\mu$m diameter, the value of β is 0.925. The optical thickness of the cloud χ_0^* needs to be large for its albedo to approach this value of β; for $\chi_0^* \simeq 125$ for instance, the albedo of this cloud $A = 0.90$. For a typical stratus cloud 0.5 km thick, $\chi_0^* \simeq 20$, giving $A \simeq 0.66$ (problem 6.9). Note in figs. 3.6 (a) and 7.9 the intense brightness of the thickest clouds.

Problems

(See the appendices for some of the physical data necessary for the solution of problems.)

6.1 Integrate the Clausius–Clapeyron equation (3.13) for water vapour between temperatures T_r and $T_\infty(T_r - T_\infty \ll T_\infty)$ to give an expression in the following form for the corresponding saturation vapour pressures $p_s(T_r)$ and $p_s(T_\infty)$

$$\ln\left[\frac{p_s(T_r)}{p_s(T_\infty)}\right] \simeq \frac{LM_r(T_r - T_\infty)}{RT_\infty^2} \qquad\qquad (6.15)$$

(R is the gas constant and M_r molecular weight of water.)

6.2 Write (6.2) in terms of vapour pressure rather than vapour density. Using (6.4) and (6.15) derive the following expression for the rate of growth of a water droplet.

$$r\frac{dr}{dt} \simeq \frac{S - 1}{[(L^2 M_r \rho_L/\lambda R T^2) + \{\rho_L R T/p_s(T_\infty)DM_r\}]} \qquad\qquad (6.16)$$

where $S = p_\infty/p_s(T_\infty)$, p_∞ being the actual vapour pressure far away from the droplet. $S - 1 (\ll 1)$ is the *supersaturation* of the vapour phase.

The effective degree of supersaturation depends also on (1) the droplet radius especially for drops $< 1\,\mu$m in radius over which the equilibrium vapour pressure is considerably higher than over a plane surface, and (2) the purity of the water, the equilibrium vapour pressure being less for salt solutions. Because of these considerations very small drops only persist on hygroscopic nuclei.

6.3 For pure water use (6.16) to calculate the time taken for a droplet to grow from 2 μm radius (1) to 10 μm radius and (2) to 40 μm radius by condensation if the supersaturation is 0.05% (i.e. $S - 1 = 5 \times 10^{-4}$) and $T = 273$ K. ($D = 0.23$ cm^2 s^{-1}.)

6.4 From Stokes' law for the viscous force F acting on a sphere of radius r falling with terminal velocity v in a fluid of viscosity η namely

$F = 6\pi\eta v r$

calculate the time taken for drops of radius 1 μm, 10 μm, 100 μm, to fall through 1 km. (η for air $= 1.7 \times 10^{-5}$ kg m^{-1} s^{-1}.)

6.5 From values of the saturation vapour pressure over plane liquid water surfaces and plane ice surfaces given in appendix 2 for $-10°$C, calculate the time taken for an ice crystal in a cloud of water droplets at $-10°$C to grow from 1 μm radius to 100 μm radius, assuming that the crystal remains spherical.

Calculate also the temperature of the ice crystal. (Latent heat of sublimation of ice $= 2800$ J g^{-1}, $D = 0.23$ cm^2 s^{-1}.)

6.6 For $\omega = 1$, find A and τ for $\chi_0^* = 20$ and $f = 0.9$. Compare with the values quoted in the chapter when $\omega = 0.9997$.

6.7 From (6.11) and (6.14) show that

$$\beta^2 = \frac{1 - \omega f - \alpha}{1 - \omega f + \alpha}$$

6.8 Derive (6.12) and (6.13).

6.9 For stratus cloud described in text with $\chi_0^* = 20$, find transmissivity τ. Also find the absorptivity of the cloud (i.e. $1 - \tau - A$).

6.10 The scattering cross-section at wavelength λ of a drop of radius r when the ratio $2\pi r/\lambda \gg 1$ is approximately $2\pi r^2$. Find the optical thickness of a cloud 0.5 km thick containing 150 drops cm^{-3} having $r = 5\,\mu$m. Remember $\chi_0^* \simeq 1.66\chi_0$.

6.11 At $2.3\,\mu m$, for drops having $r = 7\,\mu m$, $\omega = 0.988$, $f = 0.9$, find β and find the value of χ_0^* to give an albedo of 0.98β. Compare this value of χ_0^* with that for typical stratus.

6.12 At $10.6\,\mu m$, the wavelength of the CO_2 laser, $\omega = 0.36$, $f = 0.9$. Find β and the transmissivity of diffuse radiation at this wavelength in $100\,m$ of fog having the same properties as the stratus cloud in the text. (Assume χ_0^* is the same at $10.6\,\mu m$ as at $0.7\,\mu m$.)

6.13 In the visible part of the spectrum the albedo of the Venus clouds is approximately 0.9. Supposing the clouds to be very deep and that scattering is isotropic, what is the minimum value of ω for the cloud particles?

6.14 Why does the sky appear blue while clouds in general appear white?

7

Dynamics

7.1 Total and partial derivatives

So far in making simple atmospheric models we have considered energy transfer by radiation, conduction and convection. In chapter 3 the atmosphere's radiative sources and sinks were shown to be driving a thermodynamic engine which generates kinetic energy as well as potential energy. Thermodynamic considerations alone cannot tell us much more about the form of this kinetic energy. In this chapter and those that follow, therefore, we turn our attention to the atmosphere's motions, our aim being to find simple descriptions or physical models which can aid our understanding of some of the main features of the atmosphere's circulation.

When applying the laws of motion to an element of fluid, it is important to realize that the fluid being considered is in general moving with respect to our chosen frame of reference. We shall frequently need to relate the derivative with respect to time t of a scalar quantity ψ taken following the fluid, i.e. the *individual* or *total derivative* $d\psi/dt$ to the *local*, or *partial*, *derivatives* $\partial\psi/\partial t$ etc. appropriate to a fixed point in the chosen frame of reference.

Since ψ is varying in both space and time

$$d\psi = \frac{\partial\psi}{\partial t}\,dt + \frac{\partial\psi}{\partial x}\,dx + \frac{\partial\psi}{\partial y}\,dy + \frac{\partial\psi}{\partial z}\,dz$$

i.e.
$$\frac{d\psi}{dt} = \frac{\partial\psi}{\partial t} + u\frac{\partial\psi}{\partial x} + v\frac{\partial\psi}{\partial y} + w\frac{\partial\psi}{\partial z}$$

or
$$\frac{d\psi}{dt} = \frac{\partial\psi}{\partial t} + \mathbf{V}.\,\mathrm{grad}\,\psi \tag{7.1}$$

where $\quad u = \dfrac{dx}{dt}, \quad v = \dfrac{dy}{dt}, \quad w = \dfrac{dz}{dt}$

are the components of the fluid's velocity $\mathbf{V}$ with respect to Cartesian axes in our frame of reference.

Equation (7.1) will be used constantly in the paragraphs that follow.

7.2 Equations of motion

Newton's second law of motion applied to an element of fluid of density ρ moving with velocity $\mathbf{V}$ in the presence of a pressure gradient ∇p and a gravitational field (described by $\mathbf{g}'$) is (for derivation and discussion of this equation see for instance Batchelor (1967) or Rutherford (1959))

$$\frac{d\mathbf{V}}{dt} = \mathbf{g}' - \frac{1}{\rho}\nabla p + \mathbf{F} \tag{7.2}$$

where $\mathbf{F}$ is the frictional force on the element which will be discussed more fully in chapter 9.

Equation (7.2) applies to an absolute or inertial frame of reference (i.e. say fixed with respect to the solar system). Our interest is in motion relative to axes fixed with respect to the earth's surface, which is rotating with angular velocity $\mathbf{\Omega}$.

A vector $\mathbf{A}$ in frame Σ which is rotating at angular velocity $\mathbf{\Omega}$ with respect to frame Σ' will have a component of motion $\mathbf{\Omega} \wedge \mathbf{A}$ in frame Σ' due to the relative motion of the two frames (fig. 7.1) so that

$$\left(\frac{d\mathbf{A}}{dt}\right)_{\Sigma'} = \left(\frac{d\mathbf{A}}{dt}\right)_{\Sigma} + \mathbf{\Omega} \wedge \mathbf{A} \tag{7.3}$$

If we denote differentiation in the Σ' frame and quantities referred to that frame by primes, unprimed quantities being referred to the Σ frame, and let $\mathbf{A} = \mathbf{r}$ (the radius vector) and $d'\mathbf{r}/dt$ in turn we have

$$\mathbf{V}' = \mathbf{V} + \mathbf{\Omega} \wedge \mathbf{r}$$

Fig. 7.1

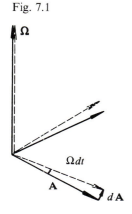

and $\dfrac{d'\mathbf{V}'}{dt} = \dfrac{d\mathbf{V}'}{dt} + \mathbf{\Omega} \wedge \mathbf{V}'$

$\qquad\qquad = \dfrac{d\mathbf{V}}{dt} + 2\mathbf{\Omega} \wedge \mathbf{V} + \mathbf{\Omega} \wedge (\mathbf{\Omega} \wedge \mathbf{r})$

so that in the rotating frame (7.2) becomes

$$\dfrac{d\mathbf{V}}{dt} + 2\mathbf{\Omega} \wedge \mathbf{V} + \mathbf{\Omega} \wedge (\mathbf{\Omega} \wedge \mathbf{r}) = -\dfrac{1}{\rho}\,\nabla p + \mathbf{g}' + \mathbf{F}$$

or $\dfrac{d\mathbf{V}}{dt} = 2\mathbf{V} \wedge \mathbf{\Omega} - \dfrac{1}{\rho}\,\nabla p + \mathbf{g} + \mathbf{F}$ (7.4)

where $\mathbf{g} = \mathbf{g}' - \mathbf{\Omega} \wedge (\mathbf{\Omega} \wedge \mathbf{r})$ (7.5)

is the acceleration due to gravity and includes as it should the centrifugal term $\mathbf{\Omega} \wedge (\mathbf{\Omega} \wedge \mathbf{r})$ (problem 7.1).

The first term on the right hand side of (7.4) is the Coriolis term which applies particularly to moving particles in rotating frames. It is perpendicular both to the direction of motion and the earth's axis of rotation. As we shall see in this and the following chapters, the Coriolis term has a profound influence on atmospheric motion.

A convenient set of axes at any point on the earth's surface (fig. 7.2) has x directed towards the east, y towards the north and z vertically upwards (or more precisely in the direction of the vector defined by (7.5)). This set is not strictly Cartesian because the directions of the axes are functions of position on the spherical earth. If u, v, w are the components of the velocity $\mathbf{V}$ in the x, y, z

Fig. 7.2

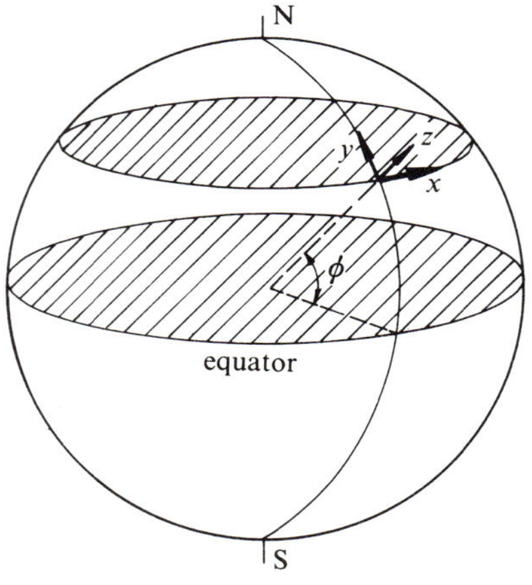

directions respectively and $\mathbf{i}, \mathbf{j}, \mathbf{k}$ are unit vectors directed along each axis then taking into account the way the axes change with position (problem 7.2)

$$\frac{d\mathbf{V}}{dt} = \left(\frac{du}{dt} - \frac{uv\tan\phi}{a} + \frac{uw}{a}\right)\mathbf{i}$$
$$+ \left(\frac{dv}{dt} + \frac{u^2\tan\phi}{a} + \frac{wv}{a}\right)\mathbf{j} + \left(\frac{dw}{dt} - \frac{u^2+v^2}{a}\right)\mathbf{k} \tag{7.6}$$

where ϕ is the latitude and a the earth's radius. Also

$$2\mathbf{V}\wedge\mathbf{\Omega} = 2\Omega(v\sin\phi - w\cos\phi)\mathbf{i}$$
$$- 2\Omega u\sin\phi\,\mathbf{j} \tag{7.7}$$
$$+ 2\Omega u\cos\phi\,\mathbf{k}$$

The magnitudes of the various terms in (7.6) and (7.7) will be very different depending on the scale of the motion under study. In this chapter we shall be concerned with motions on what is generally known as the synoptic scale or larger, that is systems of typically 1000 km in horizontal dimension, very much larger than their vertical scale (of order 1 scale height or ~ 10 km). For this scale observed vertical velocities (typically $1\,\mathrm{cm\,s^{-1}}$) are very much smaller than horizontal velocities (typically $10\,\mathrm{m\,s^{-1}}$) so that, in the momentum equation (7.4) terms involving w can, to a first approximation, be neglected. Such motion is described as *quasi-horizontal*. Because of this, and because those terms in (7.6) which involve a in the denominator are smaller by about one order of magnitude than the other terms (problem 7.3) and may therefore again to a first approximation be neglected, the equation of motion (7.4) may be simplified to become

$$\frac{d\mathbf{V}}{dt} = f\mathbf{V}\wedge\mathbf{k} - \frac{1}{\rho}\nabla p + \mathbf{F} \tag{7.8}$$

where $f = 2\Omega\sin\phi$ $\qquad\qquad\qquad\qquad$ (7.9)

and where now

$$\frac{d\mathbf{V}}{dt} = \mathbf{i}\frac{du}{dt} + \mathbf{j}\frac{dv}{dt}$$

and all the terms are vectors in the horizontal plane.

For the vertical direction, under the same approximation (problem 7.3) the hydrostatic equation (1.4) applies.

We now consider two important approximations to (7.8) in which for different situations some of the terms may be neglected in comparison to the others. Other approximations are mentioned in the problems at the end of the chapter.

7.3   The geostrophic approximation

For large scale motion away from the surface the friction **F** is small. Further, for steady flow with small curvature $dV/dt \simeq 0$. The resulting motion is known as *geostrophic*. The geostrophic velocity $\mathbf{V}_g$ is given by (fig. 7.3)

$$f\mathbf{V}_g \wedge \mathbf{k} = \frac{1}{\rho}\,\mathbf{\nabla}p \tag{7.10}$$

describing the familiar situation in which the wind blows parallel to the isobars, for the northern hemisphere in a clockwise direction around centres of high pressure (*anticyclones*) and anticlockwise around centres of low pressure (*depressions* or *cyclones*). For $\phi = 30°$, $\sin\phi = \frac{1}{2}$, $f = 7.29 \times 10^{-5}$ and for a pressure gradient of 0.2 kPa $(100$ km$)^{-1}$, the appropriate velocity for balance is 23.8 m s^{-1}.

The geostrophic approximation works well at heights above ~ 1 km (at lower levels friction becomes important) and for latitudes $> \sim 10°$, so that under these conditions the wind velocity and direction may be deduced from a chart of isobars. Note that for horizontal motion, the Coriolis term is zero at the equator.

At lower levels where friction is not negligible (cf. chapter 9), since friction acts in a direction approximately opposite to **V**, the velocity can no longer be precisely parallel to the isobars (fig. 7.4). For a high pressure centre

Fig. 7.3. The geostrophic approximation.

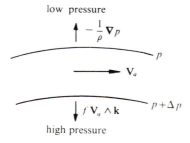

Fig. 7.4

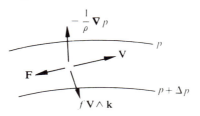

the balancing velocity has a component outwards from the centre and for a low pressure centre an inward component. To conserve the mass of air within the system, air therefore sinks over high pressure systems and rises over low pressure ones (§ 9.5), producing an important modification to the vertical velocities which would occur in the absence of friction.

7.4   Cyclostrophic motion

At low latitudes f is small and for motion having a large curvature as for instance a tropical cyclone, another approximation to (7.8) is the *cyclostrophic* one in which the acceleration of the air towards the centre (distance r away with the centre considered as the origin of co-ordinates) is balanced by the pressure gradient $\partial p/\partial r$

i.e.
$$\frac{V^2}{r} = \frac{1}{\rho}\frac{\partial p}{\partial r} \tag{7.11}$$

For a pressure gradient of 3 kPa $(100\text{ km})^{-1}$ and $r = 100$ km, $V \simeq 50$ m s^{-1} – typical values for a tropical cyclone. Note that the sense of the motion can be cyclonic or anticyclonic, but the air always blows around a region of low pressure. If both the acceleration term and the Coriolis terms are included, the resulting solution is known as the *gradient wind* (problem 7.9).

The ratio of the acceleration term dV/dt to the Coriolis term is a dimensionless number known as the *Rossby number Ro*. For motion with speed V on a scale of typical horizontal dimension L,

$$Ro \simeq \frac{V^2/L}{fV} = \frac{V}{fL} \tag{7.12}$$

In mid-latitudes a typical value of Ro is 0.1 (problem 7.7) in which case the error introduced by the geostrophic approximation is $\sim 10\%$ (problem 7.8). The smallness of the Rossby number therefore is a measure of the validity of the geostrophic approximation.

7.5   Surfaces of constant pressure

If z is the height of a surface of constant pressure p on which two neighbouring points A and B have the same y co-ordinate

$$\frac{\partial p}{\partial x}\Delta x + \frac{\partial p}{\partial z}\Delta z = \Delta p = 0$$

Since
$$\frac{\partial p}{\partial z} = -g\rho \tag{1.2}$$

we have

$$g\frac{\partial z}{\partial x} = \frac{1}{\rho}\frac{\partial p}{\partial x}$$

and similarly for

$$\partial z/\partial y$$

so that

$$g\rho\nabla_p z = \nabla_z p \tag{7.13}$$

where $\nabla_p z$ denotes the gradient of z in the surface of constant pressure. Equation (7.13) relates the gradient of pressure in a horizontal surface to the gradient of height in a nearby surface of constant pressure.

Because of the variation of g with altitude and with latitude (problem 7.11), (7.13) is not easy to use as it stands. It is convenient to define geopotential Φ as

$$\Phi = \int_0^z g\,dz$$

If now the geostrophic approximation (7.10) is written in terms of Φ, we have

$$\mathbf{V}_g = \frac{1}{f}\mathbf{k}\wedge\nabla_p\Phi \tag{7.14}$$

Notice that the density ρ has disappeared from the equation; it is therefore applicable without change to any level in the atmosphere. For this reason standard meteorological charts are plotted as contours of Φ at various pressure levels rather than as contours of pressure on horizontal surfaces. It is usual to express Φ in terms of *geopotential height* Φ/g_0, where g_0 is the standard value of g at the surface (problem 7.11).

7.6   The thermal wind equation

Horizontal pressure gradients arise in the atmosphere owing to density differences which in turn are related to temperature gradients. To relate the geostrophic wind field and its variation with height to the temperature field (7.14) can be applied to two surfaces of constant pressure p_1 and p_2 so that

$$\mathbf{V}_g(p_2) - \mathbf{V}_g(p_1) = \frac{1}{f}\mathbf{k}\wedge\nabla_p(\Phi_2 - \Phi_1) \tag{7.15}$$

Now, from the hydrostatic equation

$$\Phi_2 - \Phi_1 = \frac{R\bar{T}}{M_r}\ln\frac{p_1}{p_2} \tag{7.16}$$

where $\bar{T}$ is the mean temperature between the surfaces so that

$$\mathbf{V}_g(p_2) - \mathbf{V}_g(p_1) = \mathbf{V}_t = \frac{R}{M_r f} \ln \frac{p_1}{p_2} \mathbf{k} \wedge \mathbf{V}_p \bar{T} \tag{7.17}$$

The quantity $\mathbf{V}_t$ is called the *thermal wind* over the interval (p_1, p_2); it is related to the $\bar{T}$ field in a similar way to which $\mathbf{V}_g$ is related to the Φ field (fig. 7.5). The quantity $\Phi_2 - \Phi_1$, is known as the *thickness*; thickness charts are effectively charts of the mean temperature between different pressure levels.

Various illustrations of the use of the thermal wind are given in the problems at the end of the chapter. Fig. 5.1 shows the mean fields of zonal wind and temperature illustrating, for the atmosphere as a whole, how the two are related.

7.7　The equation of continuity

A further equation we shall require in later chapters is the equation of continuity which states that the net flow of mass into unit volume per unit time is equal to the local rate of change of density.

In the elementary volume (fig. 7.6) where the density is ρ and u, v, w the velocity components along Cartesian axes, the mass entering the volume per unit time at x over the face of area $dy\,dz$ is $\rho u\,dy\,dz$, and leaving at $x + dx$ is $(\rho u + (\partial/\partial x)(\rho u)\,dx)\,dy\,dz$. Adding all the components together, we can write,

Fig. 7.5

Fig. 7.6

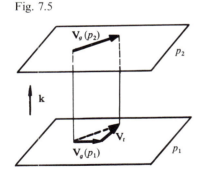

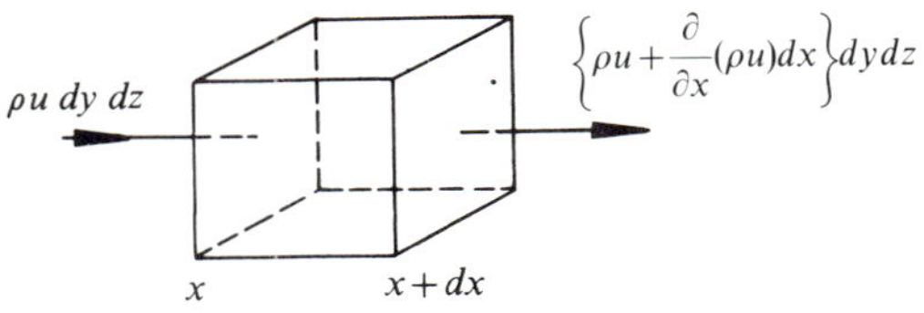

$$\frac{\partial(\rho u)}{\partial x} + \frac{\partial(\rho v)}{\partial y} + \frac{\partial(\rho w)}{\partial z} = -\frac{\partial \rho}{\partial t}$$

or $$\operatorname{div} \rho \mathbf{V} = -\frac{\partial \rho}{\partial t} \qquad (7.18)$$

Equation (7.18) is the *equation of continuity*. For an incompressible fluid it reduces to

$$\operatorname{div} \mathbf{V} = 0 \qquad (7.19)$$

A useful form of the continuity equation in which the pressure p is the vertical co-ordinate is found by considering an elemental column of atmosphere (fig. 7.7) confined between the surfaces of constant pressure p and $p - \delta p$. The mass of the column δm is equal to $\rho\, \delta x\, \delta y\, \delta z$ which by the hydrostatic equation gives

$$\delta m = \frac{\delta x\, \delta y\, \delta p}{g}$$

Following the motion, the mass δm is conserved i.e.

$$0 = \frac{1}{\delta m}\frac{d}{dt}(\delta m) = \frac{g}{\delta x\, \delta y\, \delta p}\frac{d}{dt}\left(\frac{\delta x\, \delta y\, \delta p}{g}\right)$$

Carrying out the differentiation and taking the limits we find

$$\frac{\partial u}{\partial x} + \frac{\partial v}{\partial y} + \frac{\partial \omega}{\partial p} = 0 \qquad (7.20)$$

where $\omega = dp/dt$.

Equation (7.20) is the continuity equation in isobaric co-ordinates; note that, as with many other equations written in the isobaric system, it does not involve the density.

Fig. 7.7

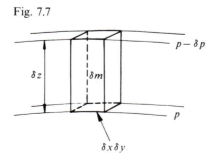

Problems

7.1 Calculate the value of the centripetal acceleration of a particle at the equator and compare with g. What is the deviation of the plumb-line direction from the true direction of the centre of a uniform spherical earth at latitude 45°?

7.2 Show that

$$\frac{d\mathbf{V}}{dt} = \mathbf{i}\,\frac{du}{dt} + \mathbf{j}\,\frac{dv}{dt} + \mathbf{k}\,\frac{dw}{dt}$$

$$+ u\,\frac{d\mathbf{i}}{dt} + v\,\frac{d\mathbf{j}}{dt} + w\,\frac{d\mathbf{k}}{dt} \tag{7.21}$$

Show that for the set of axes defined by fig. 7.2

$$\frac{d\mathbf{i}}{dt} = u\,\frac{\partial \mathbf{i}}{\partial x} = \frac{u}{a\cos\phi}\,(\mathbf{j}\sin\phi - \mathbf{k}\cos\phi) \tag{7.22}$$

$$\frac{d\mathbf{j}}{dt} = u\,\frac{\partial \mathbf{j}}{\partial x} + v\,\frac{\partial \mathbf{j}}{\partial y} = -\frac{u\tan\phi}{a}\,\mathbf{i} - \frac{v}{a}\,\mathbf{k} \tag{7.23}$$

$$\frac{d\mathbf{k}}{dt} = u\,\frac{\partial \mathbf{k}}{\partial x} + v\,\frac{\partial \mathbf{k}}{\partial y} = \frac{u}{a}\,\mathbf{i} + \frac{v}{a}\,\mathbf{j} \tag{7.24}$$

Hence derive (7.6).

7.3 For synoptic scale motions of horizontal dimension $\sim 10^3$ km, of vertical dimension ~ 10 km for which a typical horizontal velocity is $10\ \mathrm{m\,s^{-1}}$, a typical vertical velocity is $1\ \mathrm{cm\,s^{-1}}$, estimate the relative magnitudes of the various terms in (7.6) and (7.7). Compare the magnitude of the terms in the vertical direction with g.

7.4 What is the pressure gradient required at the earth's surface at 45° latitude to maintain a geostrophic wind velocity of $30\ \mathrm{m\,s^{-1}}$?

7.5 For motion around a centre of pressure 100 km away, at 30° latitude, compute the wind speed for which the Coriolis term will be equal to the acceleration term.

7.6 In the absence of a pressure gradient show that the radius of curvature of the flow is $-V/f$ and is anticyclonic. What is the period of a complete oscillation? Such flow is known as *inertial* flow and has been observed in the oceans as well as in the atmosphere.

7.7 Compute the value of the Rossby number for a typical case (latitude 45°, $L \simeq 1000$ km, $V \simeq 10\ \mathrm{m\,s^{-1}}$). For the same case, what is the value of the Rossby number on Mars and Venus?

7.8 When $Ro = 0.1$ what is the error in the geostrophic wind approximation?

7.9 When friction is neglected in (7.8) show that when the other terms are retained, in the presence of a pressure gradient $\partial p / \partial r$ for flow with radius of curvature r,

$$V = -\frac{fr}{2} \pm \left[\frac{f^2 r^2}{4} + \frac{r}{\rho} \frac{\partial p}{\partial r} \right]^{1/2} \qquad (7.25)$$

This approximation is known as the *gradient wind*. Note that r is measured from the centre of curvature and V is considered positive when cyclonic and negative when anticyclonic.

Show that for anticyclonic curvature

$$\left| \frac{\partial p}{\partial r} \right| < \frac{\rho r f^2}{4} \qquad (7.26)$$

and hence that for anticyclones the pressure gradient decreases towards the centre. This is why the pressure gradients are small and the winds light near the centre of an anticyclone.

Draw diagrams showing the balance of forces for the two cyclonic and two anticyclonic solutions of equation (7.25) and comment on their physical realisability.

7.10 The atmospheric surface pressure at radius r_0 from the centre of a tornado rotating with constant angular velocity ω is p_0. Show that the surface pressure at the centre of the tornado is

$$p_0 \exp \left(-\omega^2 r_0^2 M_r / 2RT \right)$$

where the temperature T is assumed constant.

7.11 The geopotential height z_g is defined such that

$$z_g = \frac{\Phi}{g_0} = \frac{1}{g_0} \int_0^z g \, dz \qquad (7.27)$$

where Φ is the geopotential and where g_0 has a standard value of 9.807 m s^{-2}, its mean value at the surface.

Derive an expression for the variation with height of the acceleration due to gravity. Over a place where the value of g at the surface ($z = 0$) is equal to g_0, what is the difference between z_g and the geometric height z when $z = 100 \text{ km}$?

The value of g decreases by approximately 0.5% between the equator and the poles. Again for $z = 10 \text{ km}$ and 100 km, compute the differences in geopotential height z_g and geometric height z, over the equator and over the poles (cf. appendix 4).

7.12 On the 70 kPa surface at latitude $45°$ the contours for 3300 m and 3500 m geopotential height are 1000 km apart. What is the magnitude of the geostrophic wind?

7.13 Fig. 7.8(a) is a cross-section of part of a front. At a certain level, the air possesses temperatures T_2 and T_1 respectively $(T_2 > T_1)$ on the two sides of the front which has a slope of α relative to the horizontal. Apply the hydrostatic equation and the geostrophic wind equation to the region AB, together with the condition that the pressure must be continuous across the frontal surface. Show that in equilibrium the components of velocity v_2 and v_1 in the y direction on the two sides of the front satisfy the relation

$$(T_1 - T_2)g \tan \alpha = (v_2 T_1 - v_1 T_2)f \qquad (7.28)$$

If $T_2 - T_1 = 3$ K, $v_1 - v_2 = 10$ m s^{-1}, find α. The slopes of typical fronts

Fig. 7.8. (a) Cross-section through a front. (b) Warm sector of typical depression showing warm and cold fronts.

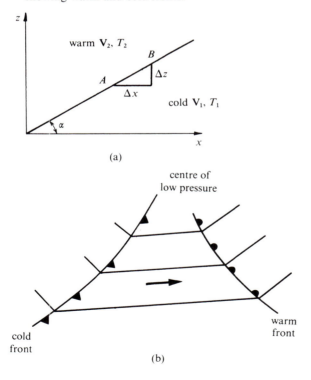

vary from around $\frac{1}{50}$ to $\frac{1}{300}$. Note that if $f=0$, i.e. the earth were not rotating, the sloping surface could not be in equilibrium.

The cloud features associated with a front are illustrated in fig. 7.9.

7.14 From the sense of the velocity change from (7.28), show that the kink in the isobars is always cyclonic (fig. 7.8(b)).

7.15 The wind at the surface is from the west. At cloud level it is from the south. Do you expect the temperature to rise or fall?

7.16 If the pole is 40 K colder than the equator and the surface wind is zero what wind would you expect at the 20 kPa pressure level? Compare the temperature and wind fields of fig. 5.1 and show so far as you can that they satisfy (7.17).

7.17 At all levels from 100 kPa to 30 kPa for the front shown in fig. 7.8(a) assume a temperature difference $T_2 - T_1$ of 5 K. For 45° latitude apply the thermal wind equation to the region between two places $x=0$ and $x=500$ km, the front being at the surface at $x=0$ and at the 30 kPa level at $x=500$ km. If the surface wind is zero what wind speed will occur at the 30 kPa level? This rather crude calculation shows that the presence of high upper level winds in the vicinity of a front roughly parallel to the frontal surface is consistent with the thermal wind equation. The concentrated region of high winds in the upper troposphere associated with fronts is known as the *jet stream.*

7.18 For a horizontal temperature gradient in the y direction only, derive a differential form of the thermal wind equation (7.17) giving the shear with height of the wind component along the x axis, i.e.

$$\frac{\partial u}{\partial z} = -\frac{g}{Tf}\frac{\partial T}{\partial y} \qquad (7.29)$$

7.19 For horizontal flow (i.e. $w=0$) show that

$$\text{div}_h\, \mathbf{V} = \mathbf{\nabla}_h \cdot \mathbf{V} = \left(\mathbf{i}\frac{\partial}{\partial x} + \mathbf{j}\frac{\partial}{\partial y}\right) \cdot (\mathbf{i}u + \mathbf{j}v)$$

Hence using some of the results of problem 7.2 show that

$$\text{div}_h\, \mathbf{V} = \frac{\partial u}{\partial x} + \frac{\partial v}{\partial y} - \frac{v}{a}\tan\phi \qquad (7.30)$$

Fig. 7.9. This very high resolution visible image over western Europe was taken by the satellite NOAA-9 at 1324 on 3 October 1985 and received at the University of Dundee Electronics Laboratory. An anticyclone with mainly clear skies was drifting away towards south-eastern Europe, while a deep depression, responsible for the swirl of cloud in the top left hand corner of the picture, was moving slowly from the Atlantic towards northern Britain. The conspicuous cloud band from the British Isles to western France and north-west Spain marks an eastward-moving cold front that separated very warm subtropical air that covered most of Europe from cool polar maritime air over western Britain and the Atlantic.

The bright (and therefore thick) band of cloud associated with the cold front over extreme western Europe results from the ascent of the warm subtropical air over the cooler Atlantic air. It is mostly layered, consisting of water droplets and ice crystals, and produced an area of rain. Embedded within the cloud mass, particularly over northern England, there are some clouds with lumpy appearance, indicating cumuliform development where the air is convectively unstable. The area of cloud ahead of the cold front over Scandinavia is formed by more widespread ascent of the subtropical air.

Immediately behind the front is a largely cloud-free zone where the cold air is subsiding. Further west, in the same cold air mass, there are widespread convective clouds producing showers. Here the polar air is colder than the sea and the air is unstable. Just to the west of Scotland is a 'comma-shaped' cluster of convective cells associated with a region of locally enhanced ascent.

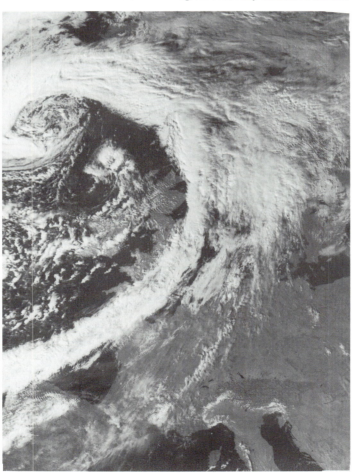

7.20 From (7.30) show that if ρ is considered constant the divergence of the geostrophic wind $\mathbf{V}_g$ is

$$\mathrm{div}_h\,\mathbf{V}_g = -\frac{v_g}{a}\,(\tan\phi + \cot\phi) \qquad (7.31)$$

7.21 For the co-ordinate system of fig. 7.2 show that

$$\mathrm{div}\,\mathbf{V} = \frac{\partial u}{\partial x} + \frac{\partial v}{\partial y} - \frac{v\tan\phi}{a} + \frac{\partial w}{\partial z} + \frac{2w}{a} \qquad (7.32)$$

Show from typical values that the last term $2w/a$ may be neglected in comparison with $\partial w/\partial z$.

7.22 From (7.10) and (7.30) calculate an expression for the divergence of the geostrophic wind. Show by putting in typical values that the terms including $(\partial/\partial x)(1/\rho)$, $(\partial/\partial y)(1/\rho)$ are small compared with the others. Also show, again by inserting typical values, that div $\mathbf{V}_g$ is an order of magnitude smaller than either of the terms $\partial u/\partial x$, $\partial v/\partial y$.

7.23 From (7.1) and (7.18) show that an alternate form of the continuity equation is

$$\frac{d\rho}{dt} + \rho\,\mathrm{div}\,\mathbf{V} = 0 \qquad (7.33)$$

7.24 Show that when the pressure p is used as the vertical co-ordinate, for a scalar ψ

$$\frac{d\psi}{dt} = \frac{\partial\psi}{\partial t} + u\frac{\partial\psi}{\partial x} + v\frac{\partial\psi}{\partial y} + \frac{dp}{dt}\cdot\frac{\partial\psi}{\partial p} \qquad (7.34)$$

7.25 Starting with (7.18) and using the hydrostatic equation derive the continuity equation in isobaric co-ordinates (7.20) by algebraic manipulation.

7.26 Omitting the frictional force $\mathbf{F}$, multiply (7.4) by ρ and take its curl to obtain the *vorticity equation*

$$\mathbf{V}\wedge\left(\rho\frac{d\mathbf{V}}{dt}\right) = \mathbf{V}\wedge(2\rho\mathbf{V}\wedge\mathbf{\Omega}) - \mathbf{g}\wedge\mathbf{V}\rho \qquad (7.35)$$

Hence show that a state of hydrostatic equilibrium (i.e. $\mathbf{V}=0$ everywhere) is impossible unless there are no variations of density on surfaces of constant geopotential. This result is known as the *Bjerknes–Jeffreys theorem*.

7.27 Venus is a planet having a very small rate of rotation about its axis. Its upper atmosphere possesses a strong zonal motion. Assuming zonal flow u, no meridional flow (i.e. $v=0$), hydrostatic equilibrium in the vertical and a negligible Coriolis term, show that the equation for cyclostrophic balance is

$$u^2 \tan \phi = -\frac{1}{\rho}\left(\frac{\partial p}{\partial \phi}\right)_z \tag{7.36}$$

p and ρ being the pressure and density at latitude ϕ and altitude z. Note that for cyclostrophic balance to hold the poleward pressure gradient must be less than zero.

Fig. 7.10. Zonal mean temperature from the surface of Venus to 100 km altitude from Pioneer Venus. Below 55 km altitude the data is from the entry probes and above that altitude from the VORTEX instrument on the orbiter. The reference pressure p_0 for the ordinate is 9216 kPa. (After Schofield & Taylor, 1983)

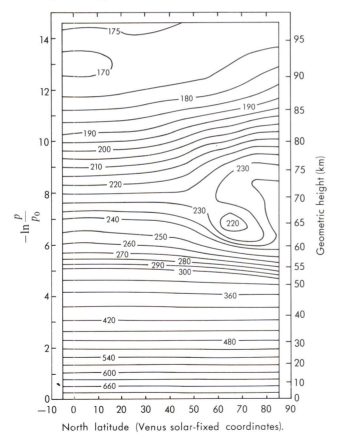

North latitude (Venus solar-fixed coordinates).

Adopting pressure as the vertical co-ordinate, show that

$$u^2(p, \phi) = -\frac{R}{M \tan \phi} \int_{p_0}^{p} \left(\frac{\partial T}{\partial \phi}\right)_p d(\ln p) + u^2(p_0, \phi)$$

From the temperature information of fig. 7.10 assuming $u(p, \phi) = 0$ at an altitude of 50 km, calculate the field of $u(p, \phi)$ for altitudes between 50 and 70 km and for ϕ between $30°$ and $65°$. What is the maximum velocity of the jet?

8

Atmospheric waves

8.1   Introduction

In chapter 7 we dealt with various approximations to the equations of motion, particularly those applicable to the fairly large scale. No attempt was made to consider how different forms of motion might arise. Before approaching a discussion of why the circulation of the atmosphere is as we find it, it is instructive to look at simple but important types of wave motion which are present in the atmosphere. To isolate simple wave forms we shall again make severe approximations.

For each case the appropriate equations are (1) the momentum equations (§ 7.2), (2) the continuity equation (§ 7.7) and (3) the first law of thermodynamics. Our method of solution will be the perturbation method.

8.2   Sound waves

To illustrate the method we first consider sound waves. For a simple treatment of these it will suffice to assume motion along the x direction only and that the y and z components and gradients in these directions are zero. The Coriolis term and friction are also omitted and the motion is assumed adiabatic. The momentum equation (7.8) becomes

$$\frac{du}{dt}+\frac{1}{\rho}\frac{\partial p}{\partial x}=0 \tag{8.1}$$

the continuity equation (7.18) becomes

$$\frac{d\rho}{dt}+\rho\frac{\partial u}{\partial x}=0 \tag{8.2}$$

and the first law of thermodynamics for adiabatic motion is

$$p\rho^{-\gamma}=constant \tag{8.3}$$

where γ is the ratio of specific heats of dry air. Combining (8.2) and (8.3) we obtain

$$\frac{1}{\gamma}\frac{d\ln p}{dt}+\frac{\partial u}{\partial x}=0 \tag{8.4}$$

The perturbation method consists of allowing the variables to be written as the sum of a mean component (represented by a bar) and a variable component (represented by a prime)

i.e. $u=\bar{u}+u'$

$$p=\bar{p}+p' \tag{8.5}$$

$\rho=\bar{\rho}+\rho'$

Here $p'\ll\bar{p}$, $\rho'\ll\bar{\rho}$ but u' is not necessarily $<\bar{u}$ as the solutions are still valid if $\bar{u}=0$. Substituting in (8.1) and (8.4), neglecting products of primed quantities, assuming that differentials of mean quantities are zero, and also using (7.1) we find

$$\left(\frac{\partial}{\partial t}+\bar{u}\frac{\partial}{\partial x}\right)u'+\frac{1}{\bar{\rho}}\frac{\partial p'}{\partial x}=0 \tag{8.6}$$

$$\left(\frac{\partial}{\partial t}+\bar{u}\frac{\partial}{\partial x}\right)p'+\gamma\bar{p}\frac{\partial u'}{\partial x}=0 \tag{8.7}$$

Eliminating u' between these equations,

$$\left(\frac{\partial}{\partial t}+\bar{u}\frac{\partial}{\partial x}\right)^2 p'-\frac{\gamma\bar{p}}{\bar{\rho}}\frac{\partial^2 p'}{\partial x^2}=0 \tag{8.8}$$

which is a wave equation having solutions

$$p'=\mathrm{Re}\left\{A\exp ik(x-ct)\right\} \tag{8.9}$$

with a constant A and

$$c=\bar{u}\pm(\gamma\bar{p}/\bar{\rho})^{1/2} \tag{8.10}$$

The speed of sound wave relative to the flow $\bar{u}$ is therefore $(\gamma\bar{p}/\bar{\rho})^{1/2}$.

8.3   Gravity waves

Since sound waves are longitudinal, only one dimension need be considered to obtain the sound wave solution. Other waves exist where the oscillation is transverse to the direction of propagation; some of these waves are known as *gravity waves*. Their existence is illustrated by the discussion of § 1.4 where we considered the stability of a parcel displaced vertically from its equilibrium level. In problem 1.10 for a stable stratification the frequency of oscillation (the Brunt–Vaisala frequency) of such a parcel was found on the assumption that the environment is unaffected by the motion – a simple calculation which applies in the limiting case of large vertical scale and small horizontal scale.

To obtain a reasonably simple solution to the basic equations for gravity waves we shall (1) work in two dimensions and ignore motion or gradients along the y direction, (2) assume that the horizontal scale of the wave is sufficiently small that the Coriolis term may be neglected by comparison with the other terms. (3) ignore the friction term, (4) assume that the unperturbed atmosphere is at rest. The components of the momentum equation (7.4) are

$$\frac{du}{dt} + \frac{1}{\rho}\frac{\partial p}{\partial x} = 0 \tag{8.11}$$

$$\frac{dw}{dt} + \frac{1}{\rho}\frac{\partial p}{\partial z} + g = 0 \tag{8.12}$$

The continuity equation (7.33) becomes

$$\frac{1}{\rho}\frac{d\rho}{dt} + \frac{\partial u}{\partial x} + \frac{\partial w}{\partial z} = 0 \tag{8.13}$$

and the first law of thermodynamics for adiabatic motion means that the potential temperature θ remains constant, i.e.

$$\frac{d\ln\theta}{dt} = 0 \tag{8.14}$$

From an equation similar to (3.3), θ can be expressed in terms of ρ and p such that

$$\frac{d\ln\theta}{dt} = \frac{1}{\gamma}\frac{d\ln p}{dt} - \frac{d\ln\rho}{dt} \tag{8.15}$$

For the unperturbed atmosphere neither θ, p nor ρ vary in the horizontal.

Let perturbations be introduced in (8.11), (8.12), (8.13) and (8.15), such that $u=u'$, $w=w'$, $\rho=\bar{\rho}+\rho'$, and $p=\bar{p}+p'$. Then using (7.1), the hydrostatic equation $\bar{p}^{-1}(\partial\bar{p}/\partial z) = -H^{-1}$, the equation of state $\bar{p}/\bar{\rho}=gH$, where H is the scale height, also neglecting products of primed quantities, the four equations (8.11), (8.12), (8.13) and (8.15) become for the unknowns u', w', $\rho'/\bar{\rho}$ and $p'/\bar{p}$ (problem 8.2):

$$\frac{\partial u'}{\partial t} + gH\frac{\partial}{\partial x}\left(\frac{p'}{\bar{p}}\right) = 0 \tag{8.16}$$

$$\frac{\partial w'}{\partial t} + g\left(\frac{\rho'}{\bar{\rho}}\right) + gH\frac{\partial}{\partial z}\left(\frac{p'}{\bar{p}}\right) - g\left(\frac{p'}{\bar{p}}\right) = 0 \tag{8.17}$$

$$\frac{\partial u'}{\partial x} + \frac{\partial w'}{\partial z} - \frac{w'}{H} + \frac{\partial}{\partial t}\left(\frac{\rho'}{\bar{\rho}}\right) = 0 \tag{8.18}$$

$$-Bw' + \frac{\partial}{\partial t}\left(\frac{\rho'}{\bar{\rho}}\right) - \frac{1}{\gamma}\frac{\partial}{\partial t}\left(\frac{p'}{\bar{p}}\right) = 0 \tag{8.19}$$

where, according to the nomenclature of problem 1.10, the vertical stability parameter B has been written for $\partial \ln \theta/\partial z$.

In solving these equations we shall assume first an isothermal atmosphere in which H is a constant, in which case $B = (\gamma - 1)/\gamma H$ and is, of course, also constant. We look for wave solutions such that each of the four unknowns vary as

$$\exp(\alpha z)\exp i(\omega t + kx + mz)$$

in which the first exponential has been introduced to allow for variation of amplitude with altitude. Solutions of this kind will exist if after substituting them into (8.16) to (8.19) the determinant of the coefficients is equal to zero. On equating imaginary parts we find that unless $m = 0$, $\alpha = 1/2H$. Solutions with the first condition have no phase variation in the vertical and are known as *external waves*; the most important of these are *surface waves* whose energy is concentrated at a boundary or discontinuity in the atmosphere similar to surface waves on the ocean. When $m \neq 0$, we have *internal waves*. Substituting $\alpha = 1/2H$ into the equation and equating the real part of the determinant of the coefficients to zero the following dispersion relation results (problem 8.3):

$$m^2 = k^2\left(\frac{\omega_B^2}{\omega^2} - 1\right) + \frac{(\omega^2 - \omega_a^2)}{c^2} \tag{8.20}$$

where c is the velocity of sound,

$$\omega_B^2 = \frac{(\gamma - 1)g}{\gamma H} \tag{8.21}$$

is the square of the Brunt–Vaisala frequency for the isothermal atmosphere (problem 8.4) and

$$\omega_a = \frac{1}{2}\left(\frac{\gamma g}{H}\right)^{1/2} = \frac{c}{2H} \tag{8.22}$$

is known as the *acoustic cut-off frequency*.

The dispersion relation (8.20) is illustrated in fig. 8.1 where curves of $m = 0$ are plotted on a ω–k diagram. Two regions where m^2 is positive are apparent, the higher frequency region ($\omega > \omega_a$ when $k = 0$) describes acoustic waves and the lower frequency region, gravity waves. Since $\omega_B < \omega_a$ (problem 8.6) these two regions are well separated. Note also from (8.20) that if $m \ll k$, i.e.

for deep waves of short horizontal wavelength compared with vertical wavelength, $\omega \simeq \omega_B$ as would be expected from our simple derivation of ω_B in problem 1.10.

If the assumption that the atmosphere is isothermal is removed, (8.20) may still be employed as a reasonable approximation with ω_B as the Brunt–Vaisala frequency for the atmosphere in question, provided $\omega_B < \omega_a$ as is normally the case throughout the atmosphere (problem 8.6).

If the atmosphere is in a state of uniform zonal flow $\bar{u}$, (8.20) still holds provided ω is replaced by $\omega + \bar{u}k$, the frequency which would be seen by an observer moving with the basic flow and known as the *intrinsic frequency*.

Equation (8.20) applies to waves having a wide range of wavelength, frequency and velocity. For gravity waves of horizontal wavelength of the order of a few kilometres, the first term is much larger than the second. In this case, allowing also for the presence of a uniform zonal wind $\bar{u}$ the dispersion relation (8.20) may be written

$$\frac{(\omega - \bar{u}k)^2}{\omega_B^2} = \frac{k^2}{m^2 + k^2} \tag{8.23}$$

The most easily observed example of such gravity waves is that of waves in the lee of mountains. Air which is forced to flow over a mountain under stable conditions is set into a gravity-wave oscillation as it moves downstream from the mountain. If the amplitude is large enough and the conditions of temperature and humidity suitable, clouds will form in the

Fig. 8.1. Dispersion curves for gravity waves. The full lines correspond to $m=0$, the dashed line is for a wave having the velocity of sound c.

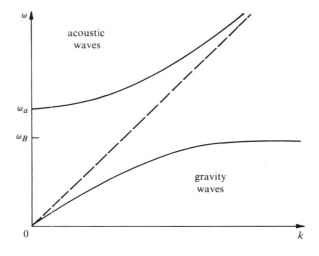

regions where the air has been lifted (figs. 3.4 and 3.5). Such waves will, of course, be stationary with respect to the mountain and the horizontal component of phase velocity relative to the surface will be zero, in which case for $k \ll m$ from (8.23) we have

$$m = \omega_B / \bar{u} \qquad (8.24)$$

Since the energy source is near the surface, energy is propagated upwards. Their group velocity is, therefore, upwards and their phase velocity downwards (problem 8.10). Consequently the lines of constant phase in the stationary waves tilt with height backwards relative to the mean flow (fig. 8.2), although in practice, because the mean wind varies with height, the situation is rarely so simple as this.

Gravity waves represent only a minor component of the motion in the lower atmosphere. Above 75 km, however, atmospheric motion is probably dominated by them (fig. 8.3). Gravity waves include tidal motions, tides being a special case of gravity waves having a particular horizontal scale and a particular period. These motions show up for instance in the variations of ionized layers as well as in winds determined from the movement of trails left by rockets or by meteors. Although mainly generated in the lower atmosphere gravity waves are propagated upwards, where as the density decreases their amplitude increases (problem 8.8). At levels above 75 km or so they are dissipated by viscous damping. Because of this they contribute significantly to the energy budget of the upper mesosphere and lower thermosphere. They also play a critical role in determining the momentum budget in the region of the mesopause (§ 10.10).

Fig. 8.2. The structure of lee waves according to Gerbier & Berenger (1961).

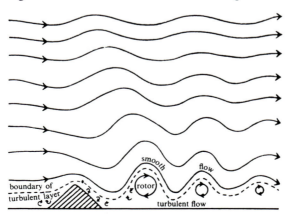

8.4 Rossby waves

The very large scale wave features which are observed in the flow on the planetary scale are known as *planetary waves* (fig. 8.4). In their simplest form they occur because of the variation of the Coriolis parameter with latitude and are known as *Rossby waves*.

The simplest Rossby wave solution is obtained for an atmosphere of constant density under the assumption of uniform zonal flow $\bar{u}$ and no vertical motion. The appropriate momentum equations (7.8) and the continuity equation (7.19) are:

$$\frac{du}{dt} + \frac{1}{\rho}\frac{\partial p}{\partial x} - fv = 0 \tag{8.25}$$

$$\frac{dv}{dt} + \frac{1}{\rho}\frac{\partial p}{\partial y} + fu = 0 \tag{8.26}$$

$$\frac{\partial u}{\partial x} + \frac{\partial v}{\partial y} = 0 \tag{8.27}$$

Operating on (8.25) with $\partial/\partial y$ and (8.26) with $\partial/\partial x$ and subtracting

$$\frac{d}{dt}\left(\frac{\partial v}{\partial x} - \frac{\partial u}{\partial y}\right) + f\left(\frac{\partial v}{\partial y} + \frac{\partial u}{\partial x}\right) + v\frac{\partial f}{\partial y} = 0 \tag{8.28}$$

Fig. 8.3. Profiles of the W–E component of the wind as measured from 36 rockets which carried chemical trail experiments all near 30°N latitude between 15 November and 14 December or near 30°S latitude between 15 May and 14 June. (From COSPAR 1972)

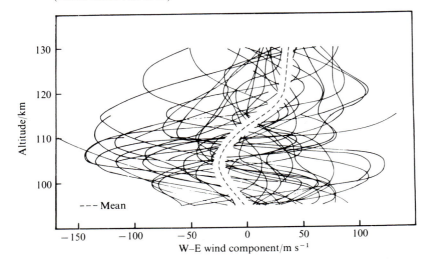

The quantity $(\partial v/\partial x) - (\partial u/\partial y)$ is one component of curl $\mathbf{V}$ which is known as the *vorticity* ζ; it may be considered as a vector equal to twice the local angular velocity of fluid elements. The second term in (8.28) is zero by (8.27). Because the Coriolis parameter f only varies with latitude it is possible to write (8.28) in the form

$$\frac{d(\zeta + f)}{dt} = 0 \tag{8.29}$$

The quantity $\zeta + f$ is known as the *absolute vorticity*; it is the vorticity due to the rotation of the fluid itself combined with that due to the earth's rotation. Equation (8.29) shows that under the conditions we have imposed of non-divergent, frictionless flow, absolute vorticity is conserved.

To solve (8.29) we assume a linear relation between f and y, i.e. we let

Fig. 8.4. Geopotential height of the 50 kPa pressure surface in decametres for a typical northern hemisphere winter situation. The large scale waves are known as *planetary waves*.

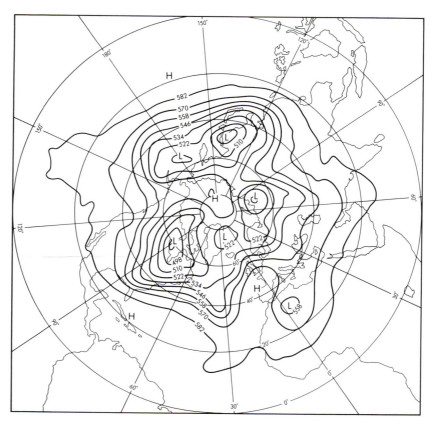

$f = f_0 + \beta y$ where β is a constant (this is known as the *β-plane approximation*). We assume a uniform zonal flow $\bar{u}$ in the unperturbed situation and introduce perturbations, i.e. $u = \bar{u} + u'$, $v = v'$ so that (8.29) becomes

$$\left(\frac{\partial}{\partial t} + \bar{u}\frac{\partial}{\partial x}\right)\left(\frac{\partial v'}{\partial x} - \frac{\partial u'}{\partial y}\right) + \beta v' = 0 \tag{8.30}$$

Because the flow is non-divergent in the horizontal a *stream function* ψ can be introduced such that by using it (8.27) is automatically satisfied,

i.e. $\qquad u' = -\frac{\partial \psi}{\partial y}, \qquad v' = \frac{\partial \psi}{\partial x} \tag{8.31}$

Substituting in (8.30),

$$\left(\frac{\partial}{\partial t} + \bar{u}\frac{\partial}{\partial x}\right)\nabla^2\psi + \beta\frac{\partial \psi}{\partial x} = 0 \tag{8.32}$$

For wave solutions $\psi = \mathrm{Re}\,\{\psi_0 \exp i(\omega t + kx + ly)\}$ to be possible the dispersion relation

$$c = -\frac{\omega}{k} = \bar{u} - \frac{\beta}{k^2 + l^2} \tag{8.33}$$

must be satisfied. The velocity relative to the zonal flow is $c - \bar{u}$ where c is the phase velocity in the x direction. Rossby waves, therefore, drift to the west relative to the basic flow, at typical speeds of a few metres per second (problem 8.14). Note that the phase speed of the waves increases with wavelength.

8.5   The vorticity equation

Under the assumptions of two dimensional flow in an atmosphere of uniform density (8.29) is a statement of the conservation of absolute vorticity. Before dealing with the propagation of Rossby-type waves in three dimensions we need to find an equation to describe vorticity changes under less stringent assumptions. For motions of synoptic scale this can be done by starting with the approximate horizontal momentum equations (8.25) and (8.26). Operating on (8.25) with $\partial/\partial y$ and (8.26) with $\partial/\partial x$ and subtracting, also noting that $df/dt = v\,\partial f/\partial y$ we have for the rate of change of absolute vorticity

$$\frac{d}{dt}(\zeta + f) = -(\zeta + f)\left(\frac{\partial u}{\partial x} + \frac{\partial v}{\partial y}\right)$$
$$-\left(\frac{\partial w}{\partial x}\frac{\partial v}{\partial z} - \frac{\partial w}{\partial y}\frac{\partial u}{\partial z}\right) + \frac{1}{\rho^2}\left(\frac{\partial \rho}{\partial x}\frac{\partial p}{\partial y} - \frac{\partial \rho}{\partial y}\frac{\partial p}{\partial x}\right) \tag{8.34}$$

Equation (8.34) is known as the *vorticity equation* (cf. (7.35)). The first term on the right hand side is the most important; it arises because of the horizontal divergence. If there is positive horizontal divergence, air is flowing out of the region in question, and the vorticity decreases. This is the same effect as occurs with a rotating body whose angular velocity decreases because of angular momentum conservation if its moment of inertia increases.

Scale analysis of (8.34) shows that for synoptic scale motions the last two terms are considerably smaller than the others (problem 8.15) and that to a first approximation

$$\frac{d_h}{dt}(\zeta + f) = -(\zeta + f)\left(\frac{\partial u}{\partial x} + \frac{\partial v}{\partial y}\right) \tag{8.35}$$

where d_h/dt denotes

$$\frac{\partial}{\partial t} + u\frac{\partial}{\partial x} + v\frac{\partial}{\partial y}$$

It is instructive to apply (8.35) to an atmosphere of constant density and temperature for which the continuity equation is (7.19) so that (8.35) becomes

$$\frac{d_h}{dt}(\zeta + f) = (\zeta + f)\frac{\partial w}{\partial z} \tag{8.36}$$

Because of the constant temperature the geostrophic wind is independent of the height z. Further, because to a first approximation the vorticity is equal to the vorticity of the geostrophic wind the vorticity will not vary with height (problem 8.16). Therefore, integrating (8.36) between levels z_1 and z_2 where $z_2 - z_1 = h$ we have

$$\frac{1}{(\zeta + f)}\frac{d_h}{dt}(\zeta + f) = \frac{w(z_2) - w(z_1)}{h} \tag{8.37}$$

Now considering the fluid which at one time is confined between the levels distance h apart, we have

$$\frac{dh}{dt} = w(z_2) - w(z_1) \tag{8.38}$$

so that (8.37) may now be written

$$\frac{d_h}{dt}\left(\frac{\zeta + f}{h}\right) = 0 \tag{8.39}$$

Equation (8.39) is a simplified statement of the conservation of *potential vorticity*. It has important consequences for atmospheric flow. Consider, for instance, adiabatic flow over a mountain barrier. As a column of air flows over the barrier its vertical extent decreases so that ζ must also decrease. A westward moving airstream will, therefore, move equatorwards as it passes over the barrier (problem 8.17).

8.6 Three dimensional Rossby-type waves

In § 8.4 we assumed two dimensional structure only for the Rossby wave solutions. To find how such waves propagate vertically we need to look for the constraints imposed on three dimensional solutions. To simplify the treatment we still wish to work with an atmosphere at nearly constant density and so we employ the *Boussinesq approximation* which allows us to neglect changes of density except where they are coupled with gravity to produce buoyancy forces. With this approximation the equation of continuity is as for an incompressible fluid, i.e. (7.19) applies, so that the appropriate vorticity equation is (8.36). With a uniform unperturbed zonal flow $\bar{u}$, the perturbation form of (8.36) is

$$\left(\frac{\partial}{\partial t}+\bar{u}\frac{\partial}{\partial x}\right)\zeta'+v'\frac{\partial f}{\partial y}-f\frac{\partial w'}{\partial z}=0 \tag{8.40}$$

where ζ has been neglected compared with f in the right hand side of (8.36) (problem 8.16).

We also require the perturbation form of the hydrostatic equation:

$$\frac{1}{\bar{\rho}}\frac{\partial p'}{\partial z}+\frac{\rho'}{\bar{\rho}}g=0 \tag{8.41}$$

and the thermodynamic equation which under the Boussinesq approximation becomes

$$\left(\frac{\partial}{\partial t}+\bar{u}\frac{\partial}{\partial x}\right)\frac{\rho'}{\bar{\rho}}-Bw'=0 \tag{8.42}$$

Comparing (8.42) with (8.19), note that since under the approximation of a nearly incompressible fluid, $\gamma \to \infty$, the last term in (8.19) may be neglected.

Eliminating ρ' and w' from (8.40), (8.41) and (8.42) and putting $\beta = \partial f/\partial y$ (the β-plane approximation) we have

$$\left(\frac{\partial}{\partial t}+\bar{u}\frac{\partial}{\partial x}\right)\left(\zeta'+\frac{f_0}{gB\bar{\rho}}\frac{\partial^2 p'}{\partial z^2}\right)+\beta v'=0 \tag{8.43}$$

Now most of the vorticity perturbation arises from the vorticity perturbation of the geostrophic wind (problem 8.16) whose components u'_g and v'_g are

$$u'_g=-\frac{1}{f_0\bar{\rho}}\frac{\partial p'}{\partial y}, \qquad v'_g=\frac{1}{f_0\bar{\rho}}\frac{\partial p'}{\partial x} \tag{8.44}$$

For the geostrophic wind, as in (8.31), we can define a stream function ψ which, comparing with (8.44), will be equal to $p'/f_0\bar{\rho}$. Introducing this into (8.43) and also approximating v' by its geostrophic value in the last term, (8.43) becomes

$$\left(\frac{\partial}{\partial t}+\bar{u}\,\frac{\partial}{\partial x}\right)\left(\nabla^2\psi+\frac{f_0^2}{gB}\,\frac{\partial^2\psi}{\partial z^2}\right)+\beta\,\frac{\partial\psi}{\partial x}=0 \tag{8.45}$$

These last substitutions involve the *quasi-geostrophic approximation* in which the wind is replaced by its geostrophic value except in the divergence term. For wave solutions

$$\psi=\mathrm{Re}\,\{\psi_0\exp i(\omega t+kx+ly+mz)\}$$

to be possible for (8.45), the dispersion relation

$$c=-\frac{\omega}{k}=\bar{u}-\frac{\beta}{(k^2+l^2+m^2f_0^2/gB)} \tag{8.46}$$

must be satisfied. Since $f_0^2/gB\ll1$ (problem 8.18), vertical wavelengths are of the order of 1% of horizontal wavelengths.

We might expect some of these large scale waves to be forced by features at the surface, by mountains or large land masses, in which case they would be stationary with respect to the surface. For c to be zero from (8.46)

$$m^2=\frac{gB}{f_0^2}\left\{\frac{\beta}{\bar{u}}-(k^2-l^2)\right\} \tag{8.47}$$

Fig. 8.5. Temperature cross-section (in K) of the stratosphere around the latitude circle 65°N as derived from the Selective Chopper Radiometer on Nimbus 4 for 4 January 1971 at the peak of a *stratospheric warming*. Note the large amplitude wavenumber one (the wavenumber refers to the number of complete cycles of a variation which occur around a latitude circle) and the westward tilt of the wave with height. (After Houghton, 1972)

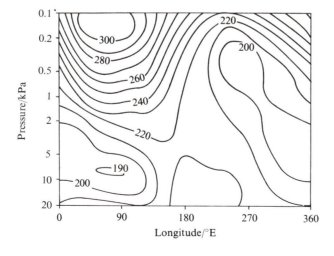

For vertical propagation to occur m^2 must be positive, which sets the condition

$$0 < \bar{u} < \beta/(k^2 + l^2) \tag{8.48}$$

Under conditions of easterly flow, therefore, or very large westerly flow no vertical propagation will occur. This result, first obtained by Charney & Drazin (1961), is confirmed by the lack of planetary wave activity in the stratosphere during the summer months when the mean stratospheric flow is easterly. Further the waves of largest horizontal wavelength propagate most readily. Figs. 8.5 and 8.6 illustrate some such waves observed in the stratosphere; where their amplitude is sufficiently large they are known as *stratospheric warmings*.

Fig. 8.6. Illustrating a quasi-stationary wave in southern hemisphere stratosphere of wavenumber 2 as observed by Selective Chopper Radiometer on Nimbus 4 satellite on 25 September 1971 (after Harwood, 1975). The quantity plotted is temperature in K of a layer ~ 10 km thick centred at (a) ~ 45 km altitude and (b) 15 km altitude. Note that there is about half a wavelength of the wave in the vertical between 45 and 15 km.

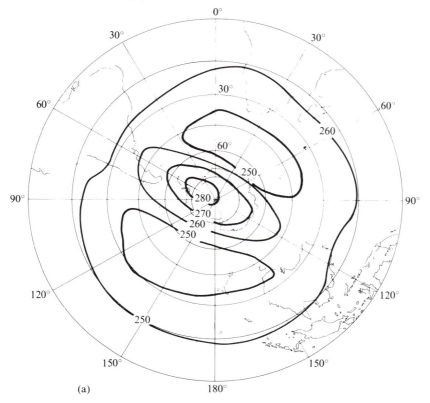

(a)

Problems

8.1 Write down the horizontal equations of motion for an atmosphere in which pressure oscillations are negligible and in which there is no basic flow, i.e.

$$\frac{\partial u}{\partial t} - fv = 0$$

$$\frac{\partial v}{\partial t} + fu = 0 \qquad (8.49)$$

Substituting $V = u + iv$ $(i = \sqrt{-1})$ solve the resulting equation on the assumption that f is constant and show that *inertial oscillations* occur with a period $2\pi/f$ (cf. also problem 7.6).

8.2 Derive (8.16) to (8.19) from (8.11) to (8.15).

8.3 Derive (8.20).

Fig. 8.6(b)

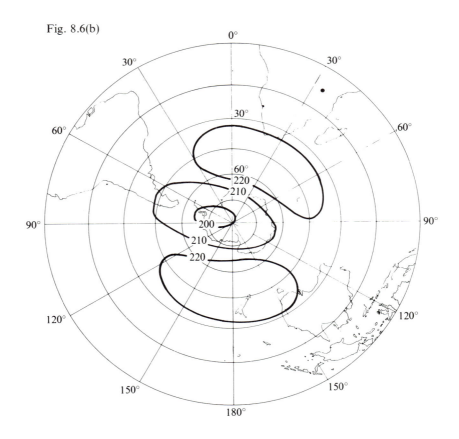

8.4 Show that for an isothermal atmosphere

$$B=(\gamma-1)/\gamma H \qquad\qquad (8.50)$$

From the result of problem 1.10 find ω_B for an isothermal atmosphere.

8.5 With $\bar{u}=0$, from (8.23), find the horizontal phase velocity of gravity waves for various values of horizontal and vertical wavelength.

8.6 Show that ω_a defined by (8.22) is always greater than ω_B for an isothermal atmosphere defined by (8.21). Calculate the value of the lapse rate of temperature with altitude for which the Brunt–Vaisala frequency is equal to ω_a.

8.7 Consider a pattern of motion in which fluid elements move at an angle θ to the vertical. Show that for an atmosphere where the Brunt–Vaisala frequency is ω_B the elements will oscillate with a frequency $\omega=\omega_B\cos\theta$.

Now (following Lindzen, 1971) imagine waves propagating upwards at angle θ to the vertical excited by a surface with uniform corrugations distant $2\pi/k$ apart moving horizontally at the base of the atmosphere. Show that the ratio of horizontal wavelength $(2\pi/k)$ to vertical wavelength $(2\pi/m)$ is $\tan\theta$. Hence deduce the relation (cf. (8.20))

$$m^2=\frac{1-(\omega/\omega_B)^2}{(\omega/\omega_B)^2}k^2$$

8.8 Show that in a wave which is propagating upwards with constant kinetic energy the velocity in the wave will vary with altitude approximately as $\exp(z/2H)$ where H is the scale height. If a gravity wave at 100 km has an amplitude of $100\ \mathrm{m\ s^{-1}}$, what would be its amplitude at the surface where it originates?

8.9 In (8.24) with $\bar{u}=10\ \mathrm{m\ s^{-1}}$, calculate the vertical wavelength for (1) an isothermal atmosphere, (2) an atmosphere with a lapse rate of $5\ \mathrm{K\ km^{-1}}$.

8.10 For gravity waves in an atmosphere with no mean flow $(\bar{u}=0)$ and having a dispersion relation given by (8.23) compute the horizontal and vertical components of the group velocity. Show that for $k\ll m$, the vertical component of group velocity is equal and opposite to the vertical component of phase velocity.

8.11 Write down the horizontal momentum equations for perturbations in an atmosphere with no mean flow $(\bar{u}=0)$, i.e.

$$\frac{\partial u'}{\partial t}+\frac{1}{\bar{\rho}}\frac{\partial p'}{\partial x}-fv'=0$$

$$\frac{\partial v'}{\partial t} + \frac{1}{\rho}\frac{\partial p'}{\partial y} + fu' = 0 \tag{8.51}$$

Add (8.17) to (8.19) in their simplified form, assume no variation with y and plane wave solutions as in § 8.3. Show that the equations are satisfied and that the appropriate dispersion relation is now

$$\omega^2 - f^2 = \frac{k^2}{m^2}(\omega_B^2 - \omega^2) \tag{8.52}$$

showing that ω is always $> f$.

This result shows the constraint on ω for gravity waves which are of sufficiently large horizontal extent for the Coriolis term to be important. It is particularly important for consideration of tides. Because f varies from zero at the equator to 2π (12 hours)$^{-1}$ at the poles, internal gravity waves with periods shorter than 12 hours can propagate anywhere but the diurnal tide will be largely confined to latitudes less than 30°.

8.12 From (8.33) calculate the phase velocity of Rossby waves relative to the basic flow for waves moving round latitude 30° of 3000 km latitudinal width and 10 000 km wavelength.

8.13 Calculate the zonal wind under which a Rossby wave pattern at 60° N of wavenumber 3 (i.e. 3 maxima around a circle of longitude) and latitudinal width 3000 km would be stationary with respect to the earth's surface.

8.14 From (8.33) find the x component of the group velocity of Rossby waves and when $k > l$ show that relative to the basic flow the group velocity is opposite to the phase velocity.

8.15 For synoptic scale motions in mid latitudes a typical value of horizontal dimension is 1000 km, of horizontal velocity 10 m s^{-1}, of vertical velocity 1 cm s^{-1}. What are corresponding typical values of fractional density variations $\delta\rho/\rho$ and pressure variations $\delta p/p$. Show that the last two terms on the right hand side of (8.34) are about an order of magnitude smaller than the first term on the right hand side.

8.16 Show by scale analysis as in problem 8.15 that in (8.35) the vorticity ζ may be approximated by the vorticity of the geostrophic wind.

Show also that under the same conditions ζ is considerably smaller than f.

8.17 Consider the implications of the theorem of conservation of potential vorticity applied to the easterly flow over a mountain barrier. A difference arises compared with westerly flow because of the variation of f with latitude. Show that for easterly flow the air must begin to move southward before it reaches the mountain barrier.

8.18 Compute values of f^2/gB (cf. (8.46)) for several typical atmospheric stabilities (e.g. for an isothermal atmosphere and for an atmosphere having a lapse rate of 5 K km^{-1}).

8.19 Compute the vertical phase velocity and group velocity for waves described by (8.46). Show that for waves in which energy is propagated upwards the phase velocity is downwards and hence that such waves slope towards the west with increasing altitude.

8.20 Write down the horizontal momentum equations (8.51) for perturbations u', v', p' in a form applicable to an equatorial β-plane, i.e. with $f = \beta y$. Assume a solution with $v' = 0$, and

$$u' = \text{Re}\,\{u'_0 \exp i(\omega t - kx)\}$$

Show that in this case the amplitude u'_0 varies with latitude as

$$\exp\left(\frac{-\beta y^2 k}{2\omega}\right)$$

Note that for a satisfactory solution $k\omega$ must be positive, i.e. the wave must be *eastward moving*. Plot the pressure field associated with the variation in zonal velocity u'. These are *Kelvin waves*; they have been observed in the equatorial stratosphere.

8.21 For gravity waves under the assumptions of §8.3 with the scale height H large compared with the vertical or horizontal wavelengths of the wave, show that the ratio of the amplitudes of the vertical to the horizontal disturbances in the wave is given by

$$\frac{|w'|}{|u'|} = \frac{-k}{m} \tag{8.53}$$

8.22 From equation (7.1) applied to the velocity $\mathbf{V}$, show, by considering individual components or otherwise, that

$$\frac{d\mathbf{V}}{dt} = \frac{\partial \mathbf{V}}{\partial t} + \boldsymbol{\zeta} \wedge \mathbf{V} + \nabla(\tfrac{1}{2}V^2) \tag{8.54}$$

where $\boldsymbol{\zeta} = \text{curl } \mathbf{V}$ is the vorticity.

From (8.54) and remembering that

$$\nabla \cdot \boldsymbol{\zeta} = 0 \tag{8.55}$$

show that

$$\mathbf{V} \wedge \frac{d\mathbf{V}}{dt} = \frac{d\zeta}{dt} - (\zeta \cdot \nabla)\mathbf{V} + \zeta(\nabla \cdot \mathbf{V}) \tag{8.56}$$

Combine (8.56) with the equation of continuity (7.33) and show that

$$\frac{1}{\rho} \mathbf{V} \wedge \frac{d\mathbf{V}}{dt} = \frac{d}{dt}\left(\frac{\zeta}{\rho}\right) - \left(\frac{\zeta}{\rho} \cdot \nabla\right)\mathbf{V} \tag{8.57}$$

Now take the equation of motion (7.2), neglecting the friction **F**, remembering that for any scalar ϕ, $\mathbf{V} \wedge \nabla\phi = 0$ and substituting in (8.57) show that

$$\frac{d}{dt}\left(\frac{\zeta}{\rho}\right) = \left(\frac{\zeta}{\rho} \cdot \nabla\right)\mathbf{V} + \left(\frac{\nabla\rho \wedge \nabla p}{\rho^3}\right) \tag{8.58}$$

A *baroclinic* situation is one in which horizontal temperature gradients exist so that pressure is not constant on surfaces of constant density or vice-versa. The last term in equation (8.58) is called the *baroclinicity vector* since it describes to what extent surfaces of constant density and surfaces of constant pressure are inclined to each other (fig. 8.7).

Take now some property which is conserved by the fluid, for instance, assuming negligible friction, the potential temperature θ, given therefore that

$$\frac{d\theta}{dt} = 0 \tag{8.59}$$

prove the relation

$$\zeta \cdot \frac{d(\nabla\theta)}{dt} = -\nabla\theta \cdot (\zeta \cdot \nabla)\mathbf{V} \tag{8.60}$$

Fig. 8.7. Illustrating *baroclinicity*. A density gradient $\nabla\rho$ exists between the heavier fluid on the left and the lighter fluid on the right. A circulation results shown by the arrows. The baroclinicity vector, which according to equation (8.25) is proportional to the rate of change of vorticity, is directed out of the paper. (After Gill, 1982)

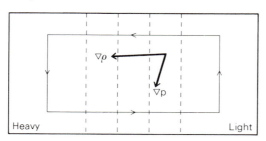

Now take the dot product of $\nabla\theta$ with (8.58) and show that

$$\frac{d}{dt}\left(\frac{\zeta}{\rho}\cdot\nabla\theta\right)=\nabla\theta\cdot\left(\frac{\nabla\rho\wedge\nabla p}{\rho^3}\right) \tag{8.61}$$

Show further that if θ is a function of p and ρ only as is the case, for instance, for a uniform composition inviscid atmosphere, then the right hand side of equation (8.61) is zero.

The quantity

$$\frac{\zeta}{\rho}\cdot\nabla\theta$$

Fig. 8.8. Isopleths of Ertel potential vorticity on an isentropic surface (potential temperature 850 K) near 30 km altitude for 7 December 1981 computed from temperature data from the Stratospheric Sounding Unit on the NOAA 7 satellite (§ 12.8). Units are $\mathrm{K\,m^2\,kg^{-1}\,s^{-1}\times10^{-4}}$. The shaded region has values between 4 and 6 units. Arrows show the geostrophic flow. The structure of the planetary wave suggests that it is 'breaking'. (After Clough *et al.*, 1985)

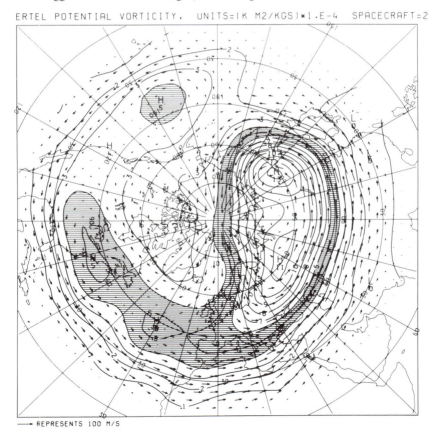

ERTEL POTENTIAL VORTICITY.   UNITS=(K M2/KGS)*1.E-4   SPACECRAFT=2

→ REPRESENTS 100 M/S

is a more general form of potential vorticity and is known as Ertel's potential vorticity. Equation (8.61) is Ertel's theorem. It states that providing there is negligible friction or dissipation, the Ertel potential vorticity is conserved following the fluid motion.

It is a very important theorem which forms the basis of much geophysical fluid dynamics (Gill, 1982 and Hoskins, McIntyre & Robertson, 1986). Fig. 8.8 illustrates the distribution of potential vorticity on an isentropic surface in the stratosphere where, because it is a conserved quantity, potential vorticity is a useful tracer of air motion.

8.23   For an atmosphere at rest the Ertel potential vorticity (equation 8.61) is approximately

$$\frac{f}{\rho} \cdot \frac{\partial \theta}{\partial z}$$

For such an atmosphere plot values of the Ertel potential vorticity as a function of height and as a function of θ for a standard March atmosphere at 10 N, 40 N and 70 N (Appendix 5), showing how its magnitude is dominated in the troposphere by the latitudinal

Fig. 8.9. Vertical cross-section from Green Bay, Wisconsin (GRB) to Apalachicola, Florida (AQQ) for 0000 GMT on 19 February 1979 showing isentropes (solid, K), geostrophic wind (dashed, m s^{-1}). The potential vorticity (dark solid, units 10^{-5} K s^{-1} kPa^{-1}) analysis is shown only for the upper portion of the frontal zone and the stratosphere. (After Uccellini *et al.*, 1985)

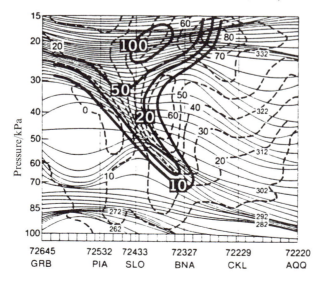

variation of f and in the stratosphere by the large values of $\partial\theta/\partial z$. Fig. 8.9 illustrates how potential vorticity can be employed to demonstrate the intrusion of stratospheric air to lower levels in the vicinity of a frontal surface.

9

Turbulence

9.1 The Reynolds Number

In writing down the equations of motion (§ 7.2) a term F was included to allow for friction. How to deal with this term is the subject of this chapter.

To begin with, consider fluid flow along a pipe. When all elements of fluid are moving along the direction of the fluid's mean motion the flow is *laminar* and internal friction in the fluid will be due only to molecular viscosity. When, however, individual elements are moving irregularly compared with the mean motion the flow is *turbulent*.

Reynold's classical experiments on the flow of fluid of uniform density through pipes showed that a non-dimensional quantity could be found whose approximate value determined whether the sheared flow was laminar or turbulent. This quantity is known as the Reynolds number Re; for a fluid of kinematic viscosity† moving with velocity V it is given by

$$Re = LV/v$$

where L is a typical length scale of the motion (e.g. the diameter of the pipe). For large values of Re greater than about 6000, turbulent flow occurs, for smaller values the flow is laminar. For air at STP $v \simeq 1.5 \times 10^{-5} \, \mathrm{m^2 \, s^{-1}}$ so that if $L = 1 \, \mathrm{m}$, provided $V > \sim 0.1 \, \mathrm{m \, s^{-1}}$ the condition that $Re > 6000$ is satisfied. With typical atmospheric scales and velocities, therefore, we expect the flow to be turbulent.

It is useful in any consideration of a particular type of atmospheric motion to restrict the consideration to a particular scale; in the last chapter this was done in order to isolate particular types of wave motion. The fact of turbulence means that this isolation can never in fact be complete. Motion on all smaller scales than the one being considered will exist; interactions and

† The kinematic viscosity *is the viscosity* divided by the density – its units are $\mathrm{m^2 \, s^{-1}}$.

exchange of energy will take place between motion on these different scales. One of the main objects, therefore, of the theory of turbulence is to provide means whereby, for any particular problem, the effect of turbulent motions on smaller scales than the one being considered can be expressed in terms of the parameters of the mean flow.

In this chapter we shall first describe the effect of turbulence in the boundary layer close to the earth's surface and then discuss in rather general terms the interactions between motion on various scales in the atmosphere at large.

9.2 Reynolds stresses

At any given point and time the components of velocity u, v and w may be expressed in terms of mean components (denoted by bars) and fluctuating components (denoted by primes), e.g. $u = \bar{u} + u'$. For this to make sense, a suitable averaging period has to be found such that the mean quantity is substantially independent of the precise averaging period. This period will depend on the particular problem under consideration; for motion in the boundary layer, 5 or 10 min is a suitable value (figs. 9.1 and 9.4).

The horizontal momentum equations (7.8) neglecting friction when due only to viscosity are

$$\frac{\partial u}{\partial t} + u \frac{\partial u}{\partial x} + v \frac{\partial u}{\partial y} + w \frac{\partial u}{\partial z} - fv + \frac{1}{\rho} \frac{\partial p}{\partial x} = 0 \tag{9.1}$$

$$\frac{\partial v}{\partial t} + u \frac{\partial v}{\partial x} + v \frac{\partial v}{\partial y} + w \frac{\partial v}{\partial z} + fu + \frac{1}{\rho} \frac{\partial p}{\partial y} = 0 \tag{9.2}$$

Fig. 9.1. Typical trace of wind speed taken at a height of 10 m over Lake Ontario.

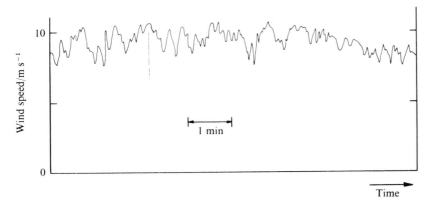

and the continuity equation (7.18) is

$$\frac{\partial}{\partial x}(\rho u)+\frac{\partial}{\partial y}(\rho v)+\frac{\partial}{\partial z}(\rho w)+\frac{\partial \rho}{\partial t}=0 \qquad (9.3)$$

Multiplying (9.1) by ρ and (9.3) by u and adding we have

$$\frac{\partial}{\partial t}(\rho u)+\frac{\partial}{\partial x}(\rho u^2)+\frac{\partial}{\partial y}(\rho uv)+\frac{\partial}{\partial z}(\rho uw)-f\rho v+\frac{\partial p}{\partial x}=0 \qquad (9.4)$$

A similar equation results from combining (9.2) and (9.3) in the same way.

We now substitute $u=\bar{u}+u'$, $v=\bar{v}+v'$ and $w=\bar{w}+w'$ into (9.4), remembering that, since $\overline{u'},\overline{v'},\overline{w'}=0$ by definition, two terms only result from averaging the product of varying quantities, e.g. $\overline{uw}=\overline{(\bar{u}+u')(\bar{w}+w')}=\bar{u}\bar{w}+\overline{u'w'}$. Since relative variations in density are very much less than those in velocity, fluctuations in the density will be neglected. The resulting equation:

$$\frac{\partial \bar{u}}{\partial t}+\bar{u}\frac{\partial \bar{u}}{\partial x}+\bar{v}\frac{\partial \bar{u}}{\partial y}+\bar{w}\frac{\partial \bar{u}}{\partial z}-f\bar{v}+\frac{1}{\rho}\frac{\partial p}{\partial x}$$
$$+\frac{1}{\rho}\left[\frac{\partial}{\partial x}\rho\overline{u'u'}+\frac{\partial}{\partial y}\rho\overline{u'v'}+\frac{\partial}{\partial z}\rho\overline{u'w'}\right]=0 \qquad (9.5)$$

contains all the terms of (9.1) now expressed in terms of the mean flow together with terms which express the effect of the fluctuating components; these terms

Fig. 9.2. Measurements of the three components of velocity and of the atmospheric temperature and of their fluctuations with time can be measured at different heights above the surface by a turbulence probe mounted on the tethering cable of a kite balloon. (Courtesy of the Meteorological Office Research Unit, Cardington)

describe what are known as the *Reynolds stresses* or the *eddy stresses*; they represent the flux of momentum due to the turbulent or eddy motions (problem 9.1).

The formulation of (9.5) suggests ways in which the Reynolds stresses might be measured (fig. 9.2), but gives no indication of how to express them in terms of the mean quantities. The simplest approach is to draw an analogy with molecular viscosity and considering a plane boundary in the xy-plane write for the eddy stress in the x direction on a plane parallel to the boundary

$$-\overline{\rho u'w'} = \rho K \frac{\partial \bar{u}}{\partial z} \tag{9.6}$$

where K is the *coefficient of eddy viscosity* (with the same dimensions as kinematic viscosity) and is effectively defined by (9.6). Typical atmospheric values of K lie in the range $1-100$ m^2 s^{-1}. These are high values when compared with the molecular viscosity of ordinary fluids (typically 10^{-5} m^2 s^{-1} for gases at STP). They approximate to the viscosity of thick treacle and demonstrate the effectiveness of eddy motions compared with molecular motions in transferring momentum.

9.3   Ekman's solution

Turbulent transfer is the main mechanism by which heat, moisture and momentum are exchanged between the atmosphere and the earth's surface. The region where these transfers occur is the *boundary layer*. First consider momentum transfer in this layer and let us assume a situation in which there is a shear of the mean wind in the vertical direction only. Neglecting, therefore, in (9.5) the stress involving horizontal gradients, we express the vertical eddy stress term (i.e. the third one) as in (9.6) with K as a constant. For a situation where the wind above the boundary layer is geostrophic and where horizontal temperature gradients through the boundary layer are negligible (i.e. the geostrophic wind through the layer does not change with height), the following equations describe the motion in the boundary layer

$$\frac{1}{\rho} \frac{\partial p}{\partial x} - fv - K \frac{\partial^2 u}{\partial z^2} = 0 \tag{9.7}$$

$$\frac{1}{\rho} \frac{\partial p}{\partial y} + fu - K \frac{\partial^2 v}{\partial z^2} = 0 \tag{9.8}$$

The components of the geostrophic wind (7.10) are

$$u_g = -\frac{1}{\rho f} \frac{\partial p}{\partial y}, \qquad v_g = \frac{1}{\rho f} \frac{\partial p}{\partial x} \tag{9.9}$$

Substituting these in (9.7) and (9.8), multiplying (9.8) by $i=\sqrt{-1}$, and adding:

$$K\frac{\partial^2(u+iv)}{\partial z^2}-if(u+iv)+if(u_g+iv_g)=0 \qquad (9.10)$$

Suitable boundary conditions are zero wind components at the surface, i.e. $u=v=0$ at $z=0$ and geostrophic wind at high levels. For simplicity, assume that this geostrophic wind is zonal, i.e. $v_g=0$, so that $u \to u_g, v \to 0$ as $z \to \infty$ is the other boundary condition.

The solution of (9.10) is

$$u+iv=u_g\left\{1-\exp\left[-\left(\frac{f}{2K}\right)^{1/2}(1+i)z\right]\right\} \qquad (9.11)$$

or in components

$$u=u_g[1-\exp(-\gamma z)\cos\gamma z] \qquad (9.12)$$

$$v=u_g\exp(-\gamma z)\sin\gamma z \qquad (9.13)$$

where $\gamma=\left(\frac{f}{2K}\right)^{1/2}$

Equations (9.12) and (9.13) describe the *Ekman spiral* (fig. 9.3). Above the level $z=\pi/\gamma$ where $v=0$ the wind is approximately geostrophic. Below this level the wind direction deviates very considerably from the geostrophic direction; at the surface for instance the deviation is $45°$. The quantity π/γ may, therefore, be considered as the approximate depth of the boundary layer. With $f=7\times10^{-5}\,\text{s}^{-1}$ and $K=10\,\text{m}^2\,\text{s}^{-1}$, $\pi/\gamma\simeq1$ km. Note that in the boundary layer the wind has a component directed generally towards low pressure, a feature which was predicted by the simple argument of §7.3.

Fig. 9.3. Wind hodograph for the lowest kilometre as measured by Dobson (1914) (dashed line) compared with Ekman spiral (full line). Figures on curves are heights in metres.

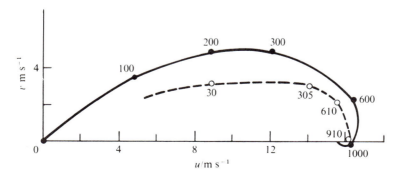

Because the approximation of constant K is not a good one, particularly near $z = 0$ (cf. fig. 9.3), the Ekman profile is not accurately followed in practice. Nevertheless as a qualitative description, the solution is a very useful one.

9.4   The mixing-length hypothesis

In the last section the eddy transfer coefficient was assumed independent of height. Especially near the surface this assumption is far from true. Some understanding of the reason for this comes from the mixing-length model of turbulence which has been valuable particularly in the near surface layers. In this model it is assumed that an element of fluid at level z moves through an average distance $|l|$ taking its velocity with it. At height $z + l$ it is reabsorbed losing all trace of its original motions. The length l therefore plays somewhat the same part as the mean free path in kinetic theory.

This description implies that

$$u' = -l \frac{\partial \bar{u}}{\partial z}$$

If the turbulence is approximately isotropic $|w'| \simeq |u'|$ and the shearing stress τ will be given by

$$\tau = -\rho \overline{u'w'} = \rho l^2 \left(\frac{\partial \bar{u}}{\partial z} \right)^2 \tag{9.14}$$

In the derivation of (9.14) it must be remembered that l is positive for upward moving parcels of fluid.

In the layer within a few tens of metres of the surface the shearing stress is approximately constant – a layer known as the *constant flux layer*. A further plausible hypothesis is that the size and path of the eddies should be proportional to height above the surface, i.e. $l = \kappa z$ where κ is known as von Karman's constant and has a value of about 0.4. On integrating (9.14) under these assumptions the wind profile is given by

$$\bar{u} = \frac{u_*}{\kappa} \ln \left(\frac{z}{z_0} \right) \tag{9.15}$$

where $u_* = (\tau/\rho)^{1/2}$ is known as the *friction velocity* and the constant of integration z_0 as the *roughness length* since it depends on the surface roughness. Equation (9.15) fits well under conditions of neutral stability. For other situations as might be expected, the wind profile and the associated momentum, heat and water vapour fluxes depend very considerably on the vertical stability.

9.5   Ekman pumping

We return to the Ekman layer and assume for the sake of simplicity that the atmosphere is of uniform density of depth H, and that in the boundary layer of depth d ($d \ll H$) the wind profile is accurately described by (9.12) and (9.13), and that above the boundary layer there is a flow u_g in the x direction, independent of height but varying with the y co-ordinate. Because of friction in the boundary layer, horizontal convergence or divergence occurs which leads through the necessity for continuity to vertical motion.

The continuity equation for a situation where density changes are neglected is (7.19)

$$\frac{\partial w}{\partial z} = -\frac{\partial u}{\partial x} - \frac{\partial v}{\partial y} \tag{9.16}$$

Substituting for u and v from (9.12) and (9.13) and integrating through the boundary layer we have for the vertical velocity w_d at $z = d$:

$$w_d = -\int_0^d \frac{\partial u_g}{\partial y} \exp(-\gamma z) \sin \gamma z \, dz \tag{9.17}$$

since $\partial u/\partial x = 0$ and since on a level surface $w = 0$ at $z = 0$.

The vorticity ζ_g of the geostrophic wind above the boundary layer is equal to $-\partial u_g/\partial y$ so that on integration (9.17) becomes

$$w_d = \tfrac{1}{2}\zeta_g \gamma^{-1} \tag{9.18}$$

For typical values (problem 9.5), w_d is a few mm s^{-1}.

The existence of a vertical velocity upwards from the boundary layer has consequences for the flow in the rest of the atmosphere, again because of continuity. Suppose for instance we consider the situation in a region of cyclonic vorticity. There is inflow in the boundary layer towards the centre of the vortex, rising air above the boundary layer and a balancing outflow at higher levels (fig. 9.4). This outflow affects the vorticity ζ_g; the rate of change of ζ_g can be found from (8.36), namely

$$\frac{\partial \zeta_g}{\partial t} = f \frac{\partial w}{\partial z} \tag{9.19}$$

Fig. 9.4. Illustrating Ekman pumping.

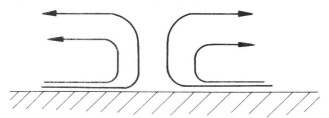

where ζ has been neglected compared with f in the right hand side of (8.36) (cf. problem 8.16). Integrating (9.19) between the top of the boundary layer ($z = d$) and the top of the atmosphere ($z = H$) we have

$$\frac{\partial \zeta_g}{\partial t}(H - d) = -fw_d$$

and on substituting from (9.18),

$$\frac{\partial \zeta_g}{\partial t} = -\frac{f}{2\gamma H}\zeta_g \qquad (9.20)$$

since $d \ll H$.

The result expressed by (9.20) is that the vorticity is reduced with a time constant of $2\gamma H/f$ – the *spin-down time* which is typically several days. The main circulation decays very much more rapidly through this means involving a *secondary circulation* than by other damping mechanisms (problem 9.5). This secondary circulation is driven by friction in the boundary layer, a mechanism known as *Ekman pumping*.

9.6   The spectrum of atmospheric turbulence

The discussion so far in this chapter has been concerned with turbulent motion on the rather small space and time scales associated with the boundary layer. Motion on much larger scales possesses random characteristics and may also be considered as turbulence which therefore occurs in the atmosphere on a very wide range of scales. Fig. 9.5 shows an estimate of the energy spectrum of atmospheric motions on different scales. The minimum in the mesoscale region corresponds to motions in size about 10 km, i.e. about 1 scale height. At smaller

Fig. 9.5. Spectral energy density of atmospheric motions having different time scales. (After Monin, 1973)

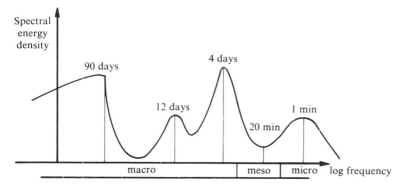

scales than this the turbulence is essentially three dimensional and not limited by the atmospheric depth, for larger scales the turbulence is approximately two dimensional. In both these regions simple dimensional arguments have been put forward in an attempt to explain the observed energy spectrum.

First suppose the existence of isotropic homogeneous three dimensional turbulence away from the atmospheric boundary. Further, suppose that in wavenumber space (wavenumber denoted by k), turbulent components at any wavenumber interact with components of similar wavenumber, and that dissipation eventually occurs at very high wavenumbers through viscosity at a rate of ε per unit mass. Under these conditions

Fig. 9.6. Spectral energy density $E(k)$, of vertical velocity at different wavenumbers k, as measured at heights $z = 1$ m and $z = 4$ m and different values of Richardson number Ri (cf. problem 9.7) shown against symbols. The line has a slope of $-\frac{5}{3}$. (From Matveev 1967)

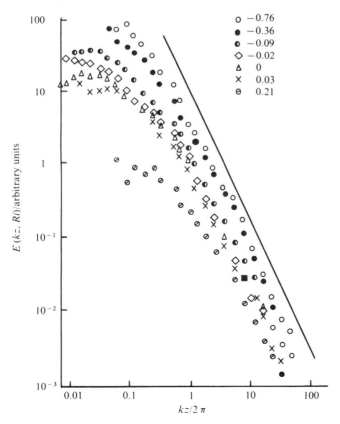

Kolmogoroff assumed that the energy spectrum $E(k)$ depends only on ε and simple dimensional considerations require (problem 9.6)

$$E(k) = \alpha_1 \varepsilon^{2/3} k^{-5/3} \tag{9.21}$$

where α_1 is a constant (cf. fig. 9.6).

Fjörtoft (1953) has shown that this energy cascade from lower to higher wavenumbers cannot occur for two dimensional turbulence. In this case it has been postulated that the energy spectrum of turbulence depends only on the rate of transfer of the mean square vorticity $\overline{\zeta^2}$, so that a dimensional argument produces (problem 9.6)

$$E(k) = \alpha_2 (\overline{\zeta^2})^{2/3} k^{-3} \tag{9.22}$$

where α_2 is a constant. Fig. 9.7 indicates that there is some evidence for this dependence.

General considerations of this kind clearly do not help an understanding of the detail of the atmospheric circulation, but they are of considerable importance in assessing the degree of interaction between motion on different scales in space and time, which assessment has a profound influence on estimates of the fundamental predictability of the general circulation.

Fig. 9.7. Kinetic energy spectra at 50 kPa, latitude 50°N for the winter season compared with power laws with exponents -2 and -3 respectively (from Julian *et al.*, 1970). Note the approximate -3 power law dependence at wavenumbers > 6.

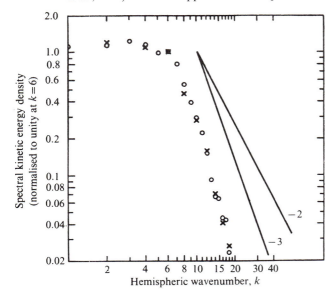

Problems

9.1 Show that the product ρuw is the instantaneous flux of u momentum in the z direction across an element of area parallel to the earth's surface where ρ, u and w are values appropriate to the position of the element of area. Show that when there is no mean vertical velocity the average momentum flux is $\rho \overline{u'w'}$.

9.2 If T' and m' represent respectively fluctuations in potential temperature and in water vapour mixing ratio, derive the following expressions for the vertical flux of heat Q and water vapour E due to turbulent motion

$$Q = \rho c_p \overline{w'T'}, \qquad E = \rho \overline{w'm'} \tag{9.23}$$

Near the surface E is also the rate of evaporation. The quantity Q/LE (where L is the latent heat of evaporation) is known as the *Bowen ratio*; it clearly depends on the moisture which is available at the surface and varies from infinity in the desert to a value of about 0.1 over the ocean.

9.3 In a similar way to (9.6), eddy transfer coefficients for heat K_Q and for water vapour K_E can be defined by

$$Q = -\rho K_Q \frac{\partial \theta}{\partial z}, \qquad E = -\rho K_E \frac{\partial m}{\partial z} \tag{9.24}$$

By considering the physical processes involved, discuss how similar the quantities K (from 9.6) K_Q and K_E may be expected to be.

9.4 In the constant flux layer the shear stress is ρu_*^2 and the heat flux Q (cf. problem 9.2). Show by dimensional analysis that a length L given by

$$L = c_p \rho T u_*^3 / \kappa g Q \tag{9.25}$$

can be formed from the quantities where von Karman's constant κ is also included. L is known as the Monin–Oboukhov length. Similarity theory postulates that vertical gradients of heat and momentum should be functions of z/L.

9.5 Put in typical values for ζ_g, K and f in (9.18) and (9.20) and work out values of w_d and of the spin-down time. Show that the spin-down time is very much shorter than the decay time which would occur under damping due to eddy viscosity but without allowing for the possibility of vertical motion and the secondary circulation.

9.6 By dimensional analysis deduce (1) the Kolmogoroff $-\frac{5}{3}$ power law equation (9.21), (2) equation (9.22).

9.7 The nature of turbulence near the surface depends on the balance between (1) the energy which can be derived from breakdown of the mean flow and (2) the working of buoyancy forces arising from the transfer of heat away from the surface.

Neglect all horizontal gradients in applying the horizontal momentum equation (9.5) to a region of the boundary layer. Leaving the acceleration and the eddy stress terms only write it as

$$\frac{\partial}{\partial t}(\rho \bar{u}) = -\frac{\partial}{\partial z}\overline{\rho u'w'} \tag{9.26}$$

Multiplying both sides of (9.26) by $\bar{u}$ derive the following expression for the rate of change of kinetic energy:

$$\frac{\partial}{\partial t}(\tfrac{1}{2}\rho\bar{u}^2) = -\frac{\partial}{\partial z}(\bar{u}\overline{\rho u'w'}) + \overline{\rho u'w'}\frac{\partial \bar{u}}{\partial z} \tag{9.27}$$

Integrate (9.27) between levels z_1 and z_2 over volume V to give

$$\frac{\partial}{\partial t}\int_V \tfrac{1}{2}\rho\bar{u}^2\,dV = -[\bar{u}\overline{\rho u'w'}]_{z_1}^{z_2} + \int_V \overline{\rho u'w'}\frac{\partial \bar{u}}{\partial z}\,dV \tag{9.28}$$

The left hand side of (9.28) is the rate of change of kinetic energy in the mean flow. On the right hand side the first term is the work done due to the drag at the boundaries z_1 and z_2 and the second term is the rate of transfer of energy from turbulence to the mean flow.

Show therefore by referring to (9.14) that the rate of destruction of turbulence per unit volume by transfer to the mean flow is $-\tau\,\partial\bar{u}/\partial z$.

Now considering the effect of buoyancy forces show that the upward force on an element of unit volume at temperature $\bar{T}+T'$ compared with the surroundings at temperature $\bar{T}$ is $T'g\rho/\bar{T}$. Hence show that the rate of work done on the eddies by buoyancy forces is $Qg/\bar{T}$ per unit volume where $Q=c_p\overline{\rho w'T'}$ (cf. problem 9.2). By comparing these expressions for the rate of work done on the eddies and for the rate of destruction of turbulence show that turbulence will persist if $Ri_F < 1$ where

$$Ri_F = -\frac{g}{\bar{T}}\frac{Q}{c_p\tau\,\partial\bar{u}/\partial z} \tag{9.29}$$

This dimensionless quantity Ri_F is known as the *flux Richardson number*. Now in (9.29) replace τ and Q by the expressions in (9.6) and (9.24) and show that turbulence will persist provided $Ri < K/K_Q$ where

$$Ri = \frac{g\,\partial\theta/\partial z}{\bar{T}(\partial\bar{u}/\partial z)^2} \tag{9.30}$$

The quantity Ri is the *Richardson number*. Under the assumption which is often made that $K \simeq K_Q$ the condition for persistent turbulence is that $Ri < 1$.

10

The general circulation

10.1 Laboratory experiments

In § 4.11 (fig. 4.5) we noticed that the atmosphere receives an excess of net radiation in equatorial regions and a deficit in polar regions. A pattern of circulation is, therefore, set up in the atmosphere which transfers heat energy from low latitudes to high latitudes.

Some insight into the type of flow to be expected in the atmosphere of a rotating planet comes from laboratory experiments on the convection which occurs in a rotating fluid. Fig. 10.1 illustrates the dependence of the type of flow on rotation rate for a laboratory experiment in which a water–glycerol solution contained in an annulus heated on the inside and cooled on the outside is rotated at different angular speeds. At low speeds the main flow is axially symmetric and heat is transferred from the centre to the outside mainly by a simple convective cell confined to the boundary layer. As the rotation rate is increased, very little transport of heat occurs unless there are significant non-axially symmetric motions (problem 10.2). Waves appear known as *baroclinic waves* rather similar to the Rossby waves discussed in § 8.4. These waves at first are regular but at higher rotation rate the preferred wavelength becomes smaller and irregularities in the wave structure occur. The main transport of heat across the annulus takes place within these waves through the mechanism of *sloping convection* (§ 10.6).

Fig. 10.2 illustrates that approximations to these simple regimes occur in the atmosphere. At latitudes near the equator the main transports occur through a meridional circulation; in mid latitudes large scale eddies are the dominant means of transport of heat and momentum.

A further understanding of the mechanisms of energy transfer comes from inspecting various terms in the budget of potential and kinetic energy as described in § 3.6 and fig. 3.3. Much of the zonal available potential energy generated by the net source of radiation at the equator and net sink over the poles is converted to eddy available potential energy and eddy kinetic energy by means of the large scale eddy motions at mid latitudes.

In chapter 7, much emphasis was placed on the importance of the geostrophic approximation for describing large-scale motion within the free atmosphere. It must also be realized that regions exist within the free atmosphere such as those close to fronts (cf. fig. 10.2) where strong gradients exist and where the flow is strongly ageostrophic. Hide (1982) has provided a

Fig. 10.1. Streak photographs illustrating the dependence of the flow type on rotation rate Ω for a laboratory 'dishpan' experiment. The values of Ω in rad s^{-1} are (a) 0.41; (b) 1.07; (c) 1.21; (d) 3.22; (e) 3.91; (f) 6.4. Working fluid was a water–glycerol solution of mean density 1.037 g cm^{-3} and kinematic viscosity 1.56×10^{-2} cm^2 s^{-1}. The streak photographs show the flow at a depth of 0.5 cm below the free upper surface (see also problem 10.1.) (From Hide & Mason, 1975)

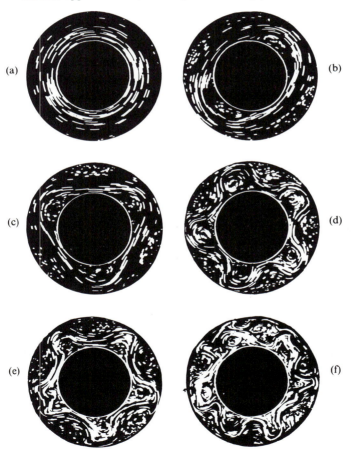

general argument as to why this should be so. Comparison between the equation for the geostrophic approximation (7.10) and the full equation of motion (7.8) shows that (7.10), being lower in order than (7.8), is mathematically degenerate. Geostrophic flow therefore cannot provide a solution for all the necessary boundary conditions: the complete solution must include all the terms of (7.8) in the analysis. This implies that (Hide 1982) regions of highly ageostrophic flow will occur not only near the boundaries of the system but also in localized regions (e.g. fronts, jet streams) of the main body of the fluid.

In the paragraphs that follow we shall look rather briefly first at a symmetric convective cell, and then at various forms of instability, especially *baroclinic instability*.

10.2   A symmetric circulation

We want first to find out whether a steady-state solution to the equations of motion can be found for an atmosphere on a rotating planet heated at the equator and cooled at the poles in which the flow shows axial symmetry, i.e. no variation with longitude.

To simplify the discussion (which partially follows Gill, 1982) the Boussinesq approximation will be assumed, i.e. changes of density are

Fig. 10.2. Schematic features of the atmospheric circulation in winter (after Palmén & Newton, 1969) showing Hadley circulation in tropics, sloping convection at mid and high latitudes; showing also polar front associated with which is polar front jet (PFJ) and the subtropical jet (STJ) associated with break in tropopause at ~ 30° latitude.

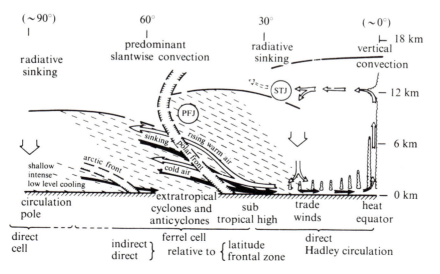

neglected except where they are coupled with gravity to produce buoyancy forces (cf. § 8.6).

We shall assume that the atmosphere is heated at a rate which varies with latitude and altitude. We also need to allow for the action of radiation processes tending to restore the atmosphere to its equilibrium situation (cf. § 2.4). The simplest way to do this is, in the thermodynamic equation, to replace $\partial/\partial t$ by $(\partial/\partial t) + \alpha$ where $1/\alpha$ is the time constant of the radiative processes (cf. § 2.4). When these terms are added to the thermodynamic equation (8.42) remembering that no variations are occurring in the x direction, we have

$$\left(\frac{\partial}{\partial t} + \alpha\right)\left(\frac{\rho'}{\rho}\right) - Bw + Q = 0 \tag{10.1}$$

where Q is the heating term and ρ' is the density perturbation associated with the circulation.

The appropriate momentum equations are (8.25) and (8.26), but, because we are looking for a steady state solution, terms describing frictional dissipation need to be added. The simplest way to do this is, in the momentum equations, to replace $\partial/\partial t$ by $\partial/\partial t + r$ where $1/r$ is the time-constant of the frictional decay process. Friction of this kind which is assumed to depend linearly on the velocity is known as *Rayleigh* friction. The momentum equations therefore become (remembering that all x variations are assumed zero)

$$\left(\frac{\partial}{\partial t} + r\right)u - fv = 0 \tag{10.2}$$

$$\left(\frac{\partial}{\partial t} + r\right)v + fu = -\frac{1}{\rho}\frac{\partial p'}{\partial y} \tag{10.3}$$

Also required is the hydrostatic equation (8.41), i.e.

$$\frac{1}{\bar{\rho}}\frac{\partial p'}{\partial z} + \frac{\rho'}{\bar{\rho}}g = 0 \tag{10.4}$$

Looking for a steady-state solution (i.e. $\partial/\partial t = 0$) and eliminating u, ρ' and p' from equations (10.1)–(10.4) we have

$$B\frac{\partial w}{\partial y} - \frac{\alpha}{rg}(f^2 + r^2)\frac{\partial v}{\partial z} - \frac{\partial Q}{\partial y} = 0 \tag{10.5}$$

The equation of continuity (7.18) for the steady-state solution is

$$\frac{\partial v}{\partial y} + \frac{\partial w}{\partial z} = 0 \tag{10.6}$$

A stream function ψ can therefore be defined by the equations

$$\frac{\partial \psi}{\partial y} = w \qquad (10.7)$$

$$\frac{\partial \psi}{\partial z} = -v \qquad (10.8)$$

which on substituting into equation (10.5) gives the equation for ψ

$$B\frac{\partial^2 \psi}{\partial y^2} + \frac{\alpha}{rg}(f^2 + r^2)\frac{\partial^2 \psi}{\partial z^2} - \frac{\partial Q}{\partial y} = 0 \qquad (10.9)$$

We consider the flow to be confined to the region $0 \leqslant y \leqslant L$, $0 \leqslant z \leqslant D$ (fig. 10.3) in which, to keep the solution simple, we shall assume the Coriolis parameter f to be constant.

A reasonable simulation of the variation of heating rate with altitude and latitude results from writing Q in the form

$$Q = Q_0 \cos ly \sin mz \qquad (10.10)$$

where $l = \pi/L$, and $m = \pi/D$

A rigid lid (i.e. $w = 0$) is assumed at the top of the atmosphere which, for this simple model, can be thought of as being at the tropopause. The stream function ψ can therefore be set to zero at $z = D$.

If a rigid surface is also assumed at the lower boundary, $\psi = 0$ at $z = 0$. In this case the solution of (10.8) is

$$\psi = \psi_0 \sin ly \sin mz \qquad (10.11)$$

Fig. 10.3. The domain for the Hadley circulation calculation.

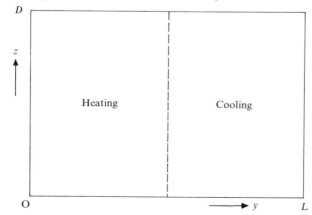

where $\psi_0 = \dfrac{lQ_0}{Bl^2 + \dfrac{\alpha m^2}{rg}(f^2 + r^2)}$ (10.12)

The solution (10.11) is plotted in fig. 10.4. Air rises in the region of maximum heating and sinks in the region of maximum cooling.

If, instead of a rigid lower boundary, it is assumed that an Ekman layer (§ 9.3) is present, much of the reverse flow is concentrated at low levels (fig. 10.4 and problems 10.5 and 10.6). Note also the easterly wind at low levels associated with the equatorward flow (cf. equation (10.2)). The resultant flow at low levels gives rise to the *trade winds* known by mariners for centuries (fig. 10.5). The circulation we have described is a *Hadley circulation* after Hadley who in 1735 described such a thermally driven cell and drew attention to the effect of the earth's rotation on air moving across circles of latitude.

It is not possible for the symmetric Hadley circulation to dominate at all latitudes. We have already mentioned in § 10.1 the importance of non-axially symmetric motions which, from the simple arguments presented there, would be expected to occur at higher latitudes. A further fundamental reason why other types of circulation must occur concerns the exchange of momentum between the solid earth and the oceans and the atmosphere. A Hadley circulation implies easterly flow at low levels, and hence transfer of easterly momentum to the surface. Since on average there can be no exchange of momentum between the atmosphere and the surface, the transfer of easterly momentum in the Hadley circulation must be balanced by a transfer of westerly momentum associated with the different flow régime at higher latitudes.

In the next few sections, we investigate the various forms of instability which occur in the atmosphere, concentrating on those that are important at high latitudes.

10.3   Inertial instability

For steady zonal flow on a rotating planet in which the geostrophic zonal velocity $\bar{u}(y)$ varies with y but not with z, consider the stability of a 'parcel' displaced with velocity v in the y direction from y_0 to $y_0 + y'$. The equations of motion describing the parcel's change in velocity are (§ 7.2)

$$\frac{du}{dt} = fv$$ (10.13)

and $$\frac{dv}{dt} = f(\bar{u} - u)$$ (10.14)

Since $v = dy/dt$, (10.13) may be integrated to give

$$u(y_0 + y') - \bar{u}(y_0) = fy' \qquad (10.15)$$

Fig. 10.4. The solution of equation (10.11) showing a Hadley circulation with rising air over the equator where the air is heated and falling air away from the equator where the air is, relatively, being cooled. The solid lines are contours of the stream function ψ with the contour interval $\frac{1}{4}$ of the maximum value. The dashed lines are contours of the x component of velocity with contour interval $\frac{1}{4}$ of the maximum speed. The dotted line is for zero zonal wind. For (a) there is no friction at the lower boundary, for (b) friction is described by problem 10.5 with the parameters ψ and λD (see problems 10.5 and 10.6) both equal to unity.

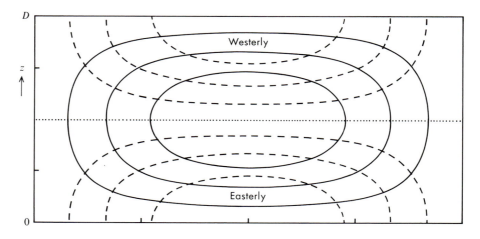

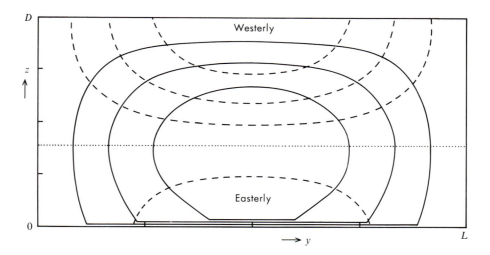

Substituting for $u(y_0 + y')$ in (10.14) we have

$$\frac{dv}{dt} = \frac{d^2 y'}{dt^2} = f\left(\frac{\partial \bar{u}}{\partial y} - f\right) y' \qquad (10.16)$$

There will therefore be stable oscillations in y' or exponential growth according as $(\partial \bar{u}/\partial y - f)$ is $<$ or >0. In the northern hemisphere with positive f the condition for *inertial stability* is therefore that $f - \partial \bar{u}/dy > 0$, i.e. that the absolute vorticity of the basic flow be positive. Observations in the atmosphere show that, on the synoptic scale, this is always the case (cf. problem 10.7).

10.4   Barotropic instability

In § 10.3, by the simple 'parcel' technique, a particular form of instability was investigated in a similar way to the instability with respect to vertical motion considered in § 1.4. There are many other forms of instability which occur in fluid motion. In this section and the next we shall consider the stability of planetary waves which were introduced in §§ 8.4 and 8.6 in two particular cases. In this section we consider a *barotropic* situation, i.e. one in which there are no horizontal temperature gradients and the basic zonal flow does not vary with height. In the next section we consider a *baroclinic* situation, i.e. one in which horizontal temperature gradients occur with a resulting shear of the zonal wind with height.

For the barotropic case the basic wave equation is the statement that absolute vorticity is conserved, i.e. (8.29):

$$\frac{d(\zeta + f)}{dt} = 0$$

Fig. 10.5. Vector mean winds over the ocean for January. (From *Meteorology for Mariners* published by the Meteorological Office)

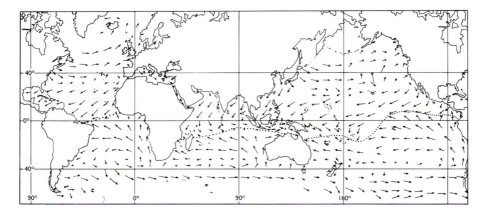

In § 8.4 in deriving the basic equation for Rossby waves, we assumed uniform zonal flow $\bar{u}$. Here we allow for a varying zonal flow with latitude so that we may find for what conditions of zonal flow instability may arise.

The perturbation form of (8.29) is

$$\left(\frac{\partial}{\partial t}+\bar{u}\frac{\partial}{\partial x}\right)\zeta'+v'\frac{\partial(\bar{\zeta}+f)}{\partial y}=0 \tag{10.17}$$

where $\dfrac{d}{dt}=\dfrac{\partial}{\partial t}+\bar{u}\dfrac{\partial}{\partial x}$

Substituting the stream function ψ defined by (8.31)

$$u'=-\frac{\partial\psi}{\partial y}, \qquad v'=\frac{\partial\psi}{\partial x}$$

we have

$$\left(\frac{\partial}{\partial t}+\bar{u}\frac{\partial}{\partial x}\right)\nabla^2\psi+\frac{\partial\psi}{\partial x}\frac{\partial(\bar{\zeta}+f)}{\partial y}=0 \tag{10.18}$$

For a wave solution

$$\psi=\mathrm{Re}\left\{\psi_0(y)\exp\left[ik(x-ct)\right]\right\} \tag{10.19}$$

where the phase velocity c may be complex, i.e.

$$c=c_r+ic_i \tag{10.20}$$

If $c_i>0$, the wave will grow exponentially with time, i.e. it will be unstable. To find in what situation this may be the case substitute from (10.19) in (10.18) so that

$$(\bar{u}-c)\left(\frac{d^2\psi_0}{dy^2}-k^2\psi_0\right)+\psi_0\frac{\partial(\bar{\zeta}+f)}{\partial y}=0 \tag{10.21}$$

To impose some boundary conditions, suppose that the wave is confined to a zonal channel of width W so that $\psi(y)=0$ at $y=0$ and $y=W$. There will now be solutions of (10.21) for certain values of the velocity c. Since the quantity ψ_0 is in general complex let us write it as $\psi_r+i\psi_i$. Multiplying (10.21) by the complex conjugate ψ_0^* of ψ_0, dividing by $\bar{u}-c$, integrating over y and equating real and imaginary parts, we have for the imaginary part

$$\int_0^W\left(\psi_i\frac{d^2\psi_r}{dy^2}-\psi_r\frac{d^2\psi_i}{dy^2}\right)dy=c_i\int_0^W\frac{\partial(\bar{\zeta}+f)}{\partial y}\frac{|\psi_0|^2}{|\bar{u}-c|^2}dy \tag{10.22}$$

The integrand on the left hand side is

$$\frac{d}{dy}\left(\psi_i\frac{d\psi_r}{dy}-\psi_r\frac{d\psi_i}{dy}\right) \tag{10.23}$$

Since $\psi_i = \psi_r = 0$ at both boundaries the integral of this expression is equal to zero. The integral on the right hand side, therefore, must be equal to zero. On inspecting the integrand we see that for $c_i > 0$, it is necessary that $\partial(\bar{\zeta} + f)/\partial y$ change sign somewhere in the region $0 < y < W$. This therefore is a necessary condition for instability. It was first worked out by Rayleigh, hence it is known as *Rayleigh's criterion*. Since $\partial\bar{\zeta}/\partial y = -\partial^2\bar{u}/\partial y^2$ and $\partial f/\partial y = \beta$ (the β-plane approximation introduced in §8.4), $\partial(\bar{\zeta} + f)/\partial y$ may be expressed as $\beta - \partial^2\bar{u}/\partial y^2$, so that the condition is that the quantity $\beta - \partial^2\bar{u}/\partial y^2$ should change sign somewhere in the region $0 < y < W$.

It is possible that this barotropic instability is responsible for the formation of tropical depressions in the *intertropical convergence zone* (ITCZ) (fig. 10.6), but it is not the form of instability responsible for the main mid-latitude disturbances.

10.5 Baroclinic instability

In the last section we considered the propagation of planetary waves under conditions of zonal flow which varied with latitude but not with altitude. Following Eady (1949) we now consider the growth of planetary waves under the same general conditions as those considered for the axially symmetric circulation in § 10.2, i.e. in a Boussinesq fluid confined between rigid horizontal planes at $z = -H/2$ and $H/2$ with a horizontal temperature gradient such that

Fig. 10.6. Variations of zonal wind and vorticity near the intertropical convergence zone illustrating the strong shear and the possibility of barotropic instability. (After Charney, 1973)

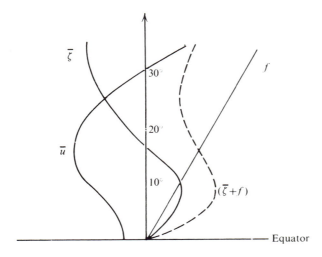

the mean zonal flow $\bar{u}$ possesses a uniform gradient with height (cf. (7.29), (10.1) and problem 10.8)

$$\frac{\partial \bar{u}}{\partial z} = -\frac{g}{f}\frac{\partial \overline{\ln \theta}}{\partial y} \tag{10.24}$$

where θ is the potential temperature. To further simplify the problem we are going to neglect the variation of Coriolis parameter with latitude, i.e. we let $\beta = 0$.

Despite this assumption we shall find that because of the baroclinic situation (i.e. the horizontal temperature gradient) planetary wave solutions exist and further that under suitable conditions instability occurs such that the waves will grow. This *baroclinic instability* is the most important form of instability in the atmosphere as it is responsible for the mid-latitude cyclones.

The relevant equations are the vorticity equation (8.45) which with $\beta = 0$ is

$$\left(\frac{\partial}{\partial t} + \bar{u}\frac{\partial}{\partial x}\right)\left(\nabla^2 \psi + \frac{f^2}{gB}\frac{\partial^2 \psi}{\partial z^2}\right) = 0 \tag{10.25}$$

and the thermodynamic equation

$$\frac{d \ln \theta}{dt} = 0$$

which on introducing the perturbations becomes

$$\left(\frac{\partial}{\partial t} + \bar{u}\frac{\partial}{\partial x}\right)(\ln \theta)' + v'\frac{\partial \overline{\ln \theta}}{\partial y} + w'\frac{\partial \overline{\ln \theta}}{\partial z} = 0 \tag{10.26}$$

The thermal wind equation (10.24) applied to the perturbation demands

$$\frac{\partial \psi}{\partial z} = \frac{g}{f}(\ln \theta)' \tag{10.27}$$

so that (10.26) becomes

$$\left(\frac{\partial}{\partial t} + \bar{u}\frac{\partial}{\partial x}\right)\frac{f}{g}\frac{\partial \psi}{\partial z} + \frac{\partial \psi}{\partial x}\frac{\partial \overline{\ln \theta}}{\partial y} + w'\frac{\partial \overline{\ln \theta}}{\partial z} = 0 \tag{10.28}$$

We look for wave-like solutions

$$\psi = \mathrm{Re}\,\{\psi_0(z)\exp i(kx + ly - kct)\} \tag{10.29}$$

which on substitution into (10.25) lead to an equation for $\psi_0(z)$, i.e.

$$\frac{f^2}{gB}\frac{\partial^2 \psi_0}{\partial z^2} - (k^2 + l^2)\psi_0 = 0 \tag{10.30}$$

which has the solution

$$\psi_0(z) = A \sinh \alpha z + C \cosh \alpha z \qquad (10.31)$$

with

$$\alpha^2 = \frac{gB}{f^2}(k^2 + l^2) \qquad (10.32)$$

Because of the rigid boundaries, boundary conditions are imposed $w' = 0$ at $z = H/2$ and $-H/2$ which on substitution from (10.31) into (10.28) lead to two simultaneous equations for A and C. These equations are only consistent if the determinant of the coefficients vanishes. An equation for c results namely (problem 10.9)

$$(c - \bar{u}_0)^2 = \left(H \frac{\partial \bar{u}}{\partial z}\right)^2 \left\{\frac{1}{4} + \frac{1}{H^2 \alpha^2} - \frac{\coth H\alpha}{H\alpha}\right\} \qquad (10.33)$$

where $\bar{u}_0$ is the value of $\bar{u}$ at $z = 0$.

When the right hand side of (10.33) is negative, $c - \bar{u}_0$ is wholly imaginary ($= c_i$). Since the equation for the stream function contains a term $\exp(kc_i t)$, under these conditions the wave amplitude will grow exponentially with a time constant $(kc_i)^{-1}$. Note that the real part of c is equal to $\bar{u}_0$ so that the phase velocity of the growing waves is equal to that of the middle of the fluid, i.e. where $z = 0$ (sometimes known as the 'steering' level).

The maximum value of kc_i occurs when $l = 0$ and $H\alpha = 1.61$ (problem 10.10 and fig. 10.7). From (10.32) for $l = 0$ the wavelength $\lambda_m = 2\pi/k$ of most

Fig. 10.7. Variation of growth rate kc_i of Eady baroclinic waves as a function of the parameter $H\alpha$ from (10.33).

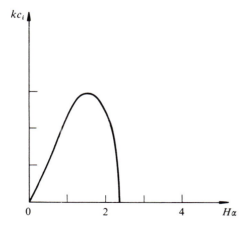

rapid growth and hence of maximum instability is given by

$$\lambda_m = \frac{2\pi H (gB)^{1/2}}{1.61f} \tag{10.34}$$

which for typical atmospheric values (problem 10.11) leads to a wavelength of ~4000 km. We should expect disturbances of about this wavelength to be the most common occurring in the atmosphere; a value which fits in well with typical wavelengths for disturbances found in mid latitudes. We further note that mid-latitude cyclones are observed to develop in periods of one to three days which agrees well with typical growth times found from the Eady theory (cf. problem 10.12).

The relative phases and amplitudes of pressure and temperature in the disturbance and also of the vertical and northward velocities are shown in fig. 10.8 (cf. problem 10.13).

10.6   Sloping convection

The baroclinic wave described in the last section is an example of *sloping convection*. Since in the wave, northward moving air is also moving upwards and southward moving air downwards (problem 10.14), the disturbance may be regarded as an eddy in which air parcels are interchanged along a slant path of average slope χ (fig. 10.9). If $\chi < \chi'$ where χ' is the slope of the isentropic surfaces in the undisturbed state, the average potential energy will decrease and kinetic energy will be released into the eddy. If $\chi > \chi'$ eddy kinetic energy will be converted to potential energy. In a rotating system where Coriolis forces dominate over the other inertial or viscous forces, sloping convection is a more effective way of transferring heat energy than axisymmetric motions of the kind considered in § 10.2. Transfer of energy by sloping convection dominates over transfer by mean meridional circulations

Fig. 10.8. Phases of temperature and pressure in an Eady wave. The motion field is also sketched in. Poleward (northward) movement is associated with rising air and southward movement with falling air.

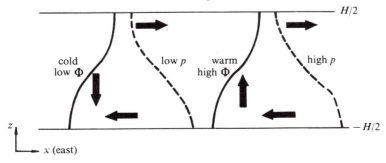

everywhere in the atmosphere outside the Hadley circulation region in the tropics.

10.7   Energy transport

For all seasons there is an excess of net radiation over the tropics and a deficit in polar regions (fig. 4.5). As a result of the atmosphere's circulation heat is transferred from equator to poles. For any given column of atmosphere (including the surface) the energy balance may be divided into a number of terms so that

$$\text{net radiation} = (\text{evaporation} - \text{precipitation}) \times \text{latent heat}$$

$$+ \text{storage} + \text{net transport by atmosphere and ocean.}$$

The net radiation is the difference between incoming solar radiation and outgoing long-wave radiation, as illustrated in fig. 4.5. Storage can be in atmosphere, land or ocean; averaged over one year the net storage will be approximately zero. Transport is either by atmospheric motions or ocean currents. Fig. 10.10 shows the northward flux of heat in the northern winter by

Fig. 10.9. Illustrating sloping convection.

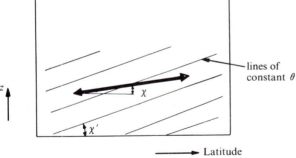

Fig. 10.10. Northward flux of heat in northern hemisphere winter by (a) mean circulations such as the tropical Hadley cell, (b) eddies. Curve (c) is the total flux. (After Newton, 1970)

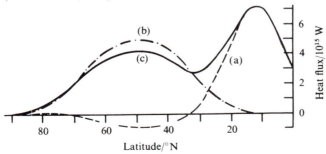

meridional motions and by eddy transfer showing that the former are effective in the tropics and the latter at higher latitudes. Heat flux due to ocean currents is difficult to estimate; it is probable comparable to that due to the atmosphere. Fig. 10.11 shows various components of the energy budget for different parts of the atmosphere, again in northern winter.

Fig. 3.3 shows more detail of the main route of the energy transformations. Zonal available potential energy generated mainly by the gross differences between incoming and outgoing radiation, is converted through the eddy motions to eddy available potential energy which in turn is converted into eddy kinetic energy which is then dissipated.

In the troposphere heat is transported by atmospheric motions down the temperature gradient from the net heat source near the equator to the heat sinks around the poles. In the lower stratosphere the eddy transport of heat is still poleward but reference to fig. 5.1 will show that, because the equatorial stratosphere is colder than the polar stratosphere, this transfer is now against the temperature gradient. This can only happen if the poleward moving air parcels sink and therefore heat adiabaticaly while the equatorward moving

Fig. 10.11. Heat and angular momentum budgets in northern hemisphere winter. In (a), units are averages in W m^{-2} for whole tropical region 5°S–32°N, and for whole extratropical region. Interior boxes give net atmospheric radiation R_a and condensation heat release LC in layers above and below 50 kPa. Dashed arrows, eddy flux e; solid arrows, cell flux c; Q_S, sensible heat flux from earth's surface. For comparison, evaporative flux Q_E is shown in parentheses. (b) shows transports of angular momentum across boundaries, and surface torques. (From Newton 1970)

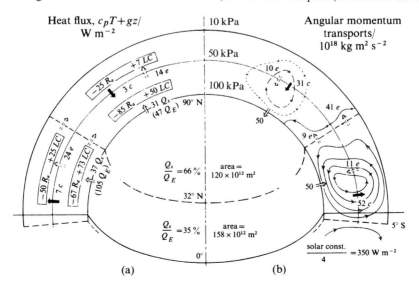

ones rise and therefore cool. Such heat transport is characteristic of a refrigerator rather than a direct heat engine, the stratospheric refrigerator being driven by energy released by the tropospheric heat engine. A diagram similar to fig. 3.3 but for the stratosphere is shown in fig. 10.12, which illustrates energy transformations from the eddy kinetic energy supplied by tropospheric forcing.

10.8　Transport of angular momentum

Apart from small effects due to tidal friction the angular momentum of the earth and its atmosphere remains constant. Since the earth's average angular velocity is very close to being constant, the atmosphere's angular momentum must also be substantially constant (problem 10.16). This means that the average transfer of angular momentum between earth and atmosphere must be close to zero. A poleward transfer of angular momentum must, therefore, occur between the tropical regions where the surface winds are easterly to mid latitudes where they are westerly. In the tropical regions this

Fig. 10.12. Average storages and main conversions of available potential energy E_A and kinetic energy E_K for 5–1 kPa layer in stratosphere as estimated by Crane (1977) from satellite data for period 28 January–10 February 1973. Units of energy in kJ m^{-2} and of conversion in W m^{-2}. Subscripts Z and E are for zonal and eddy components respectively. Compare with fig. 3.3 for the whole atmosphere.

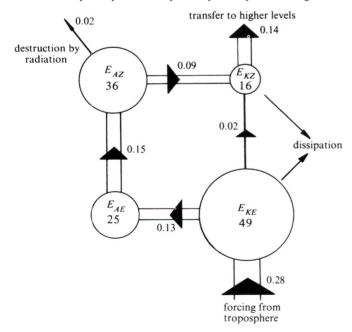

Fig. 10.13. (a) Mean angular velocity about polar axis of atmosphere relative to earth's surface; and mean annual northward angular momentum transport by (b) mean circulation, and (c) by eddies. Note the up-gradient transport of angular momentum by eddies. (After Starr, 1968)

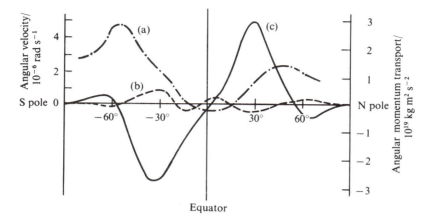

Fig. 10.14. Schematic representation of planetary waves in tropospheric flow. Note the tilting of the waves required to provide a northward momentum flux. Note also the depressions (L) and anticyclones (H) associated with the mean upper tropospheric flow (full line). (After Mintz, 1961)

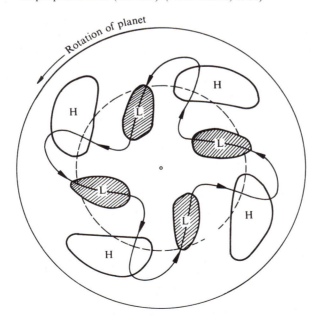

transfer of angular momentum occurs both through the Hadley circulation and through eddy transfer. In mid latitudes the transfer occurs almost entirely through the large scale eddies (figs. 10.11 and 10.13). The torque on the surface is partially due to friction in the boundary layer and partially due to systematic pressure differences between the western and eastern sides of mountain ranges especially in the northern hemisphere.

For poleward transport of angular momentum in the eddies the quantity $\overline{u'v'}$ must be positive, i.e. there must be a significant correlation between positive values of u' and v'. In a symmetrical wave, for instance the fastest growing Eady wave of § 10.5 (problem 10.17), this is not the case and the

Fig. 10.15. A spiral cloud formation seen in the atmosphere of the planet Mars taken from the Viking orbiter in 1977. The circulation which is about 200 km across is similar to that found in extra-tropical cyclones on earth. Temperatures measured simultaneously suggest that the cloud is formed of water ice. (Courtesy of NASA)

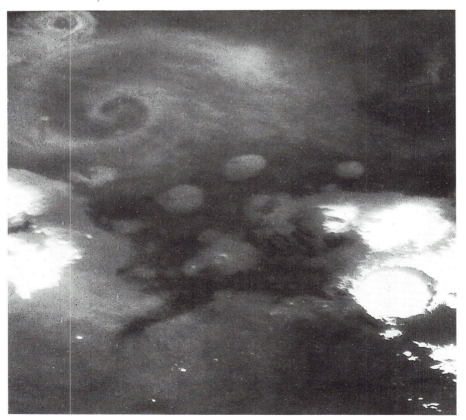

net transport will be zero. If, however, the troughs and ridges in the wave are tilted from south-west to south-east (fig. 10.14) $u' > 0$ when $v' > 0$ and northward eddy transfer of angular momentum will occur.

Notice from fig. 10.13 that much of the zonal momentum transport by eddies occurs against the gradient, thus exhibiting an effective viscosity which is negative. Similar *negative viscosity* situations in which the momentum of eddies is fed into zonal flow have been observed in atmospheres other than that of the earth for instance in the atmospheres of the sun (Starr, 1968), Mars (Leovy, 1969 and fig. 10.15) and Jupiter (Hide, 1974).

10.9 The general circulation of the middle atmosphere

So far in this chapter we have mostly concentrated on the dynamic processes which dominate the circulation of the lower atmosphere. We now consider the circulation of the middle atmosphere – the region between about 10 and 100 km in altitude. It was noted in § 10.7 and fig. 10.12 that transport in

Fig. 10.16. Cross-section of the atmosphere showing radiative equilibrium temperatures for the solstices in the stratosphere and mesosphere and convective conditions below. (After Murgatroyd, 1969)

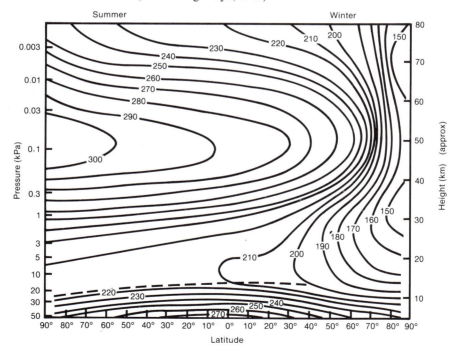

this region is largely driven from the troposphere below. Some different considerations apply, therefore, for the dynamics of the middle atmosphere.

Let us begin by asking what the temperature structure of the middle atmosphere would be if it were in radiative equilibrium. Fig. 10.16 shows this; it should be compared with fig. 5.1 where the actual temperature structure is plotted. As might be expected, the winter stratosphere and mesosphere are very much warmer than would occur under radiative equilibrium; the summer mesopause, on the other hand is very much colder.

The variation of temperature at the mesopause is just the reverse of what would be expected on grounds of radiative equilibrium. This is accomplished through an interhemispheric circulation (fig. 10.17) first suggested by Murgatroyd and Singleton (1961) and Leovy (1964). Rising air in the summer mesosphere is cooled as it ascends and is balanced by descending air in the winter mesosphere with its associated warming (problem 10.18). The upper part of the cell is completed by flow between the hemispheres in the region of the mesopause (fig. 10.17). It is easy to see in general terms how such a circulation can be associated with the thermal imbalance near the stratopause level. It is not so easy to see however what mechanism acts as the brake on the circulation. A clue to this comes from inspection of the mean zonal wind pattern shown in fig. 5.1(c). It will be noticed that the magnitude of the zonal wind relative to the earth's surface is about zero near 90 km – the mesopause level. Exchange of momentum between the surface and the mesopause which occurs through breaking gravity waves (fig. 10.18 and problem 10.19) operates in such a way so as to maintain this region of very small zonal flow (Houghton, 1978). The overall structure of the mesosphere, therefore, is determined by radiative exchange together with a circulation which, in turn satisfies (i) the thermodynamic equation (ii) the thermal wind relationship (cf. §7.6 and fig. 5.1(c)) and (iii) the requirement for very small zonal flow in the region of the mesopause (problem 10.20).

Turning now to the stratosphere we note that while the summer stratosphere is not very far removed from radiative equilibrium, the winter stratosphere is very much warmer than it would be under the assumption of radiative equilibrium. We also note that the stratosphere throughout possesses a highly stable stratification with values of potential temperature far higher than is typical in the troposphere. Vertical motion is therefore severely inhibited (problem 10.18). Clearly, however, in the winter hemisphere heat is being transported from low to high latitudes by dynamical processes.

Further clues regarding the nature of the circulation come from studies of the movement of minor constituents which act as tracers. Two of these were

Fig. 10.17. Illustrating the inter-hemispheric circulation in the mesosphere.

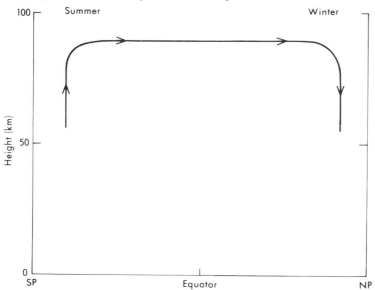

Fig. 10.18. Illustrating the behaviour of a gravity wave as it propagates upward into the mesosphere. When at level z_s the amplitude of the horizontal velocity of the wave reaches $|\bar{u}-c|$, c being the phase velocity of the wave and $\bar{u}$ the velocity of the mean flow, saturation of the wave is said to occur and momentum transfer from the wave to the mean flow begins to take place, until the critical level z_c is reached where $\bar{u}-c=0$ (see problem 10.19). (After Fritts, 1984)

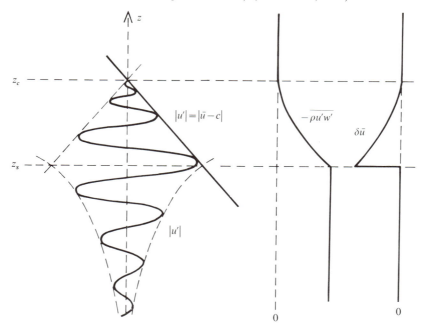

mentioned in § 5.5, namely water-vapour and ozone. Very dry air which has entered the stratosphere in the region of the tropical tropopause (problem 5.18) is carried polewards through the rest of the stratosphere, all stratospheric air being found to have a water-vapour content appropriate to saturated air at a temperature near to that of the tropical tropopause. Ozone-rich air from the tropical stratosphere is also, especially in the winter, carried polewards and downwards. A simple meridional circulation cell (Brewer, 1949) could explain this transport. Such a circulation by itself would not, however, conserve momentum nor would it necessarily transport an adequate amount of heat during the winter months.

Examples of planetary waves in the stratosphere were presented in § 8.6. There it was pointed out that planetary waves are inhibited from propagating upwards during the regime of easterly flow which occurs in the summer, but that waves of low wavenumber (i.e. large wavelength) readily propagate upwards from the troposphere into the stratosphere during the westerly flow in the winter.

The presence of a wave does not, however, imply that transport is occurring. A steady wave motion moves air about, but unless there are dissipative processes occurring or unless the amplitude of the wave is changing with time, the air will on average be unaltered by the wave motion. Dissipation occurs through radiative damping or through interaction with smaller scale motions; an example of a large 'breaking' wave where the latter are particularly important is shown in fig. 8.8. It is these dissipative and transient processes which enable the wave to interact with the mean motions and which control the transport of heat, material and momentum by the wave (Plumb (1982), World Meteorological Organisation (1986), Andrews, Holton & Leovy (1987)).

Measurements such as those made from Nimbus satellites and to be made from the Upper Atmospheric Research Satellite (§ 12.5), together with studies employing two dimensional and three-dimensional models (chapter 11) are providing the means to elucidate the complex interactions between waves and mean flow and between the processes determining the radiation transfer, the chemical transformations and the dynamics – research in the whole area being organized through the Middle Atmosphere Programme (MAP).

Problems

10.1 Consider an annulus rotating at angular velocity Ω containing liquid as shown in fig. 10.1. The equation of state relating density ρ to temperature T for the liquid is

$$\rho = \rho_0[1 - \alpha(T - T_0)]$$

where α is the thermal expansion coefficient ($\simeq 2 \times 10^{-4} \, \mathrm{K}^{-1}$). Derive the thermal wind relationship (cf. §7.6 and problem 7.18) for the annulus in the form

$$\frac{\partial \mathbf{V}}{\partial z} = \frac{\alpha g}{2\Omega} \, \mathbf{k} \wedge \nabla T \tag{10.35}$$

where $\mathbf{V}$ is the fluid's velocity.

If a Rossby number is defined for the annulus (whose interior and exterior radii are a and b respectively) by $Ro = u/\Omega(b-a)$, where u is a typical relative zonal velocity of the fluid, show by substituting in (10.35) that

$$Ro \simeq \frac{\alpha g H \Delta T}{2\Omega^2 (b-a)^2} \tag{10.36}$$

where H is the depth of the fluid and ΔT the difference in temperature across it. Estimate Ro for the situation of fig. 10.1 (a), (c) and (e) respectively for which $\Delta T = 9 \, \mathrm{K}$, $b - a = 5 \, \mathrm{cm}$, $H = 14 \, \mathrm{cm}$ and $\Omega = 0.41$, 1.21 and 3.91 rad s^{-1} respectively.

10.2 For a laboratory system of the kind described in § 10.1 with axially symmetric boundary conditions, with cylindrical co-ordinates r, ϕ, z and with the angular velocity of rotation being parallel, to the z axis, show that the radial velocity u_r can be written

$$u_r = -\frac{1}{2\rho\Omega_r} \left(\frac{\partial p}{\partial \phi} \right) + A \tag{10.37}$$

where ρ is the density, p the pressure and where the first term on the right hand side is the geostrophic term; A is the sum of all the ageostrophic terms.

Hence show that the rate of transport by advection of any quantity Q (per unit volume), such as heat, across a cylindrical surface of radius r, is given by

$$H = \int_{z_1}^{z_2} \int_0^{2\pi} u_r Q r \, d\phi \, dz$$
$$= \int_{z_1}^{z_2} \int_0^{2\pi} \left[-\frac{1}{2\rho\Omega} \left(\frac{\partial p}{\partial \phi} \right) + r A \right] Q \, d\phi \, dz \tag{10.38}$$

Experiments show that the contribution A to equation (10.38) decreases rapidly with increasing Ω. The transport H therefore will be very small unless the flow pattern departs significantly from axial symmetry. It would be expected that the flow would arrange itself in such a way that the transport of heat, for instance, would be nearer to a maximum rather than to a minimum. Even, therefore, when the boundary conditions are symmetric, departures from axial symmetric flow are to be expected (for further details of the argument see Hide (1982)).

10.3 Derive equations (10.5 and (10.9).

10.4 Show that (10.11) is a solution of (10.9).

10.5 Consider in the simple model circulation of § 10.2 the presence of a boundary layer such that the lower boundary condition for the vertical velocity w is given by equation (9.18). Show now that equation (10.9) and the boundary conditions can be satisfied by the solution for the stream function

$$\psi = \psi_0 \sin ly[\sin mz + \gamma \sinh \lambda(D - z)] \tag{10.39}$$

where $\quad \lambda^2 = \dfrac{l^2 g B}{\alpha(f^2 + r^2)} \tag{10.40}$

and where $\quad \gamma = \dfrac{m}{\beta \sinh \gamma D + \lambda \cosh \lambda D} \tag{10.41}$

with $\quad \beta = r\left(\dfrac{2}{Kf}\right)^{1/2} \tag{10.42}$

where K is defined in equation (9.6).

10.6 Insert in equations (10.40) and (10.41) values of the parameters appropriate to a Hadley circulation at tropical latitudes. Show that the values of $\gamma = 1$ and $\lambda D = 1$ used to prepare fig. 10.4 are reasonable ones.

10.7 Devise a field of zonal flow for which the absolute vorticity is <0.

10.8 Write (7.29) in terms of potential temperature and hence deduce (10.24).

10.9 Derive (10.33).

10.10 From (10.33) show that the maximum value of $H\alpha$ for which $c_i \geqslant 0$ is the solution to the equation

$$\frac{x}{2} = \coth \frac{x}{2}$$

[use the identity $\tanh x = (2 \tanh x/2)/(1 + \tanh^2 x/2)$].

10.11 Insert typical values into (10.34) and deduce values for λ_m.

10.12 With values of λ_m as found from problem 10.9, with $H\alpha = 1.61$ and with a typical value of $\partial \bar{u}/\partial z$, from (10.33) find values for $(kc_i)^{-1}$ the time constant of the rate of growth of the disturbances of greatest instability.

10.13 Substitute from (10.31) into (10.28) putting in the boundary condition $w' = 0$ at $z = \pm H/2$. Hence find the ratio C/A and show that for the wave of maximum growth the ratio is imaginary. Hence check the relative phases of quantities in the wave shown in fig. 10.8.

10.14 Show that, in the baroclinic wave of § 10.5, rising air is moving northwards and sinking air southwards.

10.15 Air from the equator is moved slowly northwards while its angular momentum is conserved. What zonal velocity will it have at latitude $30°\,N$ relative to the earth's surface?

10.16 Consider the atmosphere at rest relative to the earth. What is the ratio of its angular momentum to the angular momentum of the solid earth (assume of mean density $5500\,kg\,m^{-3}$)? Assuming constant zonal velocity for the atmosphere what variation in zonal velocity is required to produce the observed seasonal variation of 2 parts in 10^8 in the earth's rotation rate?

10.17 Show that for the fastest growing Eady wave of § 10.5, since $l = 0$, $\overline{u'v'} = 0$, so that the wave is symmetrical and no north–south momentum transport occurs.

10.18 Work out values of the potential temperature at $10\,km$ intervals at altitudes between $10\,km$ and $50\,km$ for the stratosphere at $40°\,N$ in March (cf. Appendix 5).

   Suppose stratospheric air is cooling by radiation at a rate of $0.1\,K\,day^{-1}$. How long will it take for the air to fall by $2\,km$?

10.19 Refer to fig. 10.18 where the behaviour of a gravity wave is illustrated as it propagates upwards into the mesosphere. We want to calculate the vertical flux of horizontal momentum. From equation (9.5), for a basically isothermal atmosphere with mean flow $\bar{u}(z)$ this flux is given by

$$\bar{\rho}\frac{\partial \bar{u}}{\partial t} = -\frac{\partial}{\partial z}\,\overline{\bar{\rho}u'w'} \tag{10.43}$$

Consider a gravity wave with velocities varying as (cf. §8.3)

$$u' = Re\left\{\hat{u}(z)\exp\left[\frac{z}{2H} + i(\omega t + kx + mz)\right]\right\} \tag{10.44}$$

with a similar expression for w'. The wave is said to be *saturated* when

$$\hat{u}\exp\left(\frac{z}{2H}\right) = |\bar{u} - c| \tag{10.45}$$

The result of problem 8.21 shows that, provided $m \gg 1/2H$, the vertical perturbation velocity amplitude satisfies

$$\hat{w} = -\frac{k\hat{u}}{m} \tag{10.46}$$

From equation (8.23) show that, m provided $k \ll m$,

$$m = \frac{\omega_B}{(\bar{u} - c)} \tag{10.47}$$

Using (10.47) and substituting for $\hat{u}$ and $\hat{w}$ in equation (10.43) show now that in a region where $\bar{u}$ varies substantially more slowly with height than does the density,

$$\frac{\partial \bar{u}}{\partial t} = -\frac{k(\bar{u} - c)^3}{2\omega_B H} \tag{10.48}$$

Note that the sign of the momentum exchange is such that the tendency is always to accelerate the mean flow towards the phase speed of the gravity wave.

10.20 A simple model of the interhemispheric circulation in the mesosphere, illustrated in fig. 10.17, can be set up as follows. Take the region of the mesosphere with $z = 0$ at an altitude of 50 km and $z = D$ at 85 km altitude. Take $y = 0$ at the equator and $y = \pi L$ at the pole. At the base of the region assume the zonal wind $u = -u_0 \sin(y/L)$ (approximates to the summer hemisphere of fig. 5.1 c). Assume no variation of any quantities with x. At the top of the region near the mesopause, the boundary conditions are that u is constrained to be zero by gravity wave breaking (problem 10.19). This means that an appropriate meridional velocity at the mesopause can be chosen to satisfy continuity. Within the region assume that the thermal wind equation (7.29) applies, i.e.

$$\frac{\partial u}{\partial z} = -\beta \frac{\partial T}{\partial y} \tag{10.49}$$

where β will be considered constant. Also assume that away from the boundaries the meridional velocity $v = 0$, so that to satisfy continuity the vertical velocity must be given by

$$w = w_0(y) \exp(z/H) \tag{10.50}$$

where H is the scale height which for this simple model is assumed constant. Assume the y-dependence of the temperature is governed solely by the adiabatic heating or cooling due to vertical motion, i.e.

that

$$T - T_0 = -\alpha w \tag{10.51}$$

T_0 is a mean temperature which can vary with z but not y and α is assumed independent of y or z.

Derive expressions for the variation of u and w with y and z. Note that an exactly matching circulation can be set up in the winter hemisphere. Estimate suitable values for the constants H, β and u_0, hence sketch out the fields of w, u and T.

11

Numerical modelling

11.1 A barotropic model

In previous chapters the basic equations describing atmospheric structure and motion have been solved for particular situations in order to isolate a number of specific phenomena. In the general case we need to integrate the basic equations with respect to time starting with a given atmospheric situation at a particular time so that a simulation can be provided of the atmospheric behaviour at subsequent times. The task of writing the equations and the boundary conditions in a suitable form and then of solving them with high speed digital computers is known as *numerical modelling*. By comparing the behaviour of the model with that of the real atmosphere the validity of the procedures employed by the model are tested. The most important application of numerical modelling is the development of methods sufficiently reliable and sufficiently fast to be used in routine weather forecasting. The first attempt to build such a model was made by L. F. Richardson in 1918; his book (see bibliography) is still a classic in the field. The first successful forecast was made by Charney, Fjörtoft and von Neumann (1950).

The equation employed for this early forecast was the vorticity equation applied to a barotropic atmosphere, i.e. an atmosphere which is assumed to be homogeneous and of uniform density and in which vertical motion is ignored. Writing the components of the velocity $\mathbf{V}$ in terms of the stream function ψ as defined by (8.31), the vorticity equation (8.29) becomes

$$\frac{\partial}{\partial t}(\nabla^2 \psi) + \mathbf{V} \cdot \mathbf{V}(\nabla^2 \psi + f) = 0 \tag{11.1}$$

Equation (11.1) is a one parameter equation which may be integrated numerically as it stands. It will apply to a level in a real atmosphere if a level can be found where the horizontal divergence of the flow is negligible and where coupling with other levels or with the boundaries may also be neglected. Synoptic flow in the mid troposphere is sufficiently non-divergent that for short periods (11.1) may be employed with some success.

11.2   Baroclinic models

With a barotropic model it is not possible to predict the development of instabilities due to thermal gradients such as were described in § 10.5. For these to be present in the model we must describe the motion at more than one atmospheric level, to allow for vertical motion and to use the thermodynamic equation.

Under the assumptions that the synoptic scale motions to be described are quasi-horizontal and quasi-geostrophic, a suitable form of the vorticity equation is (8.35) which on using the continuity equation (7.20) and the stream function as defined by (8.31) can be written in isobaric co-ordinates as

$$\frac{\partial}{\partial t}(\nabla^2\psi) + \mathbf{V}\cdot\mathbf{V}(\nabla^2\psi + f) - f\frac{\partial\omega}{\partial p} = 0 \tag{11.2}$$

where in writing the last term ζ has been neglected in comparison with f (problem 8.16).

The thermodynamic equation in isobaric co-ordinates is (using (7.34)):

$$\frac{\partial\ln\theta}{\partial t} + u\frac{\partial\ln\theta}{\partial x} + v\frac{\partial\ln\theta}{\partial y} + \omega\frac{\partial\ln\theta}{\partial p} = \frac{1}{c_p}\frac{dS}{dt} \tag{11.3}$$

where dS/dt is the rate of increase of entropy due to adiabatic processes (§ 11.6ff).

Now

$$d\ln\theta = \frac{1}{\gamma}d\ln p + d\ln\rho^{-1} \tag{11.4}$$

from (8.15), and the hydrostatic equation (1.2) when written in terms of the geopotential Φ (problem 7.11) gives

$$\rho^{-1} = -\partial\Phi/\partial p \tag{11.5}$$

Since the first three terms of (11.3) involve differentiation at constant pressure, with the help of (11.4) and substituting from (11.5), (11.3) becomes

$$\frac{\partial}{\partial t}\left(-\frac{\partial\Phi}{\partial p}\right) + u\frac{\partial}{\partial x}\left(-\frac{\partial\Phi}{\partial p}\right) + v\frac{\partial}{\partial y}\left(-\frac{\partial\Phi}{\partial p}\right) - \frac{\omega B}{g\rho^2} = \frac{1}{\rho c_p}\frac{dS}{dt} \tag{11.6}$$

where the static stability prameter

$$B = \frac{\partial\ln\theta}{\partial z} = -g\rho\frac{\partial\ln\theta}{\partial p} \tag{11.7}$$

The geostrophic approximation (7.14) enables the geostrophic stream function ψ defined by (8.31) to be identified with Φ/f (cf. (8.44)) so that (11.6) may be written as

$$\frac{\partial}{\partial t}\left(\frac{\partial\psi}{\partial p}\right) + \mathbf{V}\cdot\nabla\left(\frac{\partial\psi}{\partial p}\right) + \frac{\omega B}{fg\bar{\rho}^2} = -\frac{1}{\bar{\rho}c_p f}\frac{dS}{dt} \tag{11.8}$$

where in the heating term and the expression for static stability an average density $\bar{\rho}$ has been included.

The vorticity equation (11.2) and the thermodynamic equation (11.8) are the basic equations for numerical integration. To see how this is carried out for a two-level model consider the atmosphere divided up into layers as shown in fig. 11.1 with the levels denoted by subscripts 0 to 4. The equations are solved for stream functions ψ_1 and ψ_3 at the levels 1 and 3. At the top of the atmosphere the boundary condition is $\omega_0 = 0$ and similarly in the absence of orograpy $\omega_4 = 0$ at the bottom. In terms of finite differences

$$\left(\frac{\partial\omega}{\partial p}\right)_1 \simeq \frac{\omega_2}{\Delta p} \quad \text{and} \quad \left(\frac{\partial\omega}{\partial p}\right)_3 \simeq -\frac{\omega_2}{\Delta p} \tag{11.9}$$

where Δp is a pressure difference of half an atmosphere. Vorticity equations as (11.2) may, therefore, be written for levels 1 and 3, namely

$$\frac{\partial}{\partial t}\nabla^2\psi_1 + \mathbf{V}_1\cdot\nabla(\nabla^2\psi_1 + f) - f\omega_2/\Delta p = 0$$

$$\frac{\partial}{\partial t}\nabla^2\psi_3 + \mathbf{V}_3\cdot\nabla(\nabla^2\psi_3 + f) + f\omega_2/\Delta p = 0 \tag{11.10}$$

where $\mathbf{V}_1 = \mathbf{k}\wedge\nabla\psi_1$ and $\mathbf{V}_3 = \mathbf{k}\wedge\nabla\psi_3$

The finite difference form of (11.8) in terms of ψ_1 and ψ_3 and ω_2 only will be

$$\frac{\partial}{\partial t}(\psi_1 - \psi_3) + \mathbf{V}_2\cdot\nabla(\psi_1 - \psi_3) - \frac{B\Delta p}{gf\bar{\rho}^2}\omega_2 = \frac{1}{c_p}\frac{\Delta p}{\bar{\rho}f}\frac{dS}{dt} \tag{11.11}$$

where $\mathbf{V}_2 = \mathbf{k}\wedge\nabla\tfrac{1}{2}(\psi_1 + \psi_3)$

Fig. 11.1. Arrangement of levels and variables for a two-parameter baroclinic model.

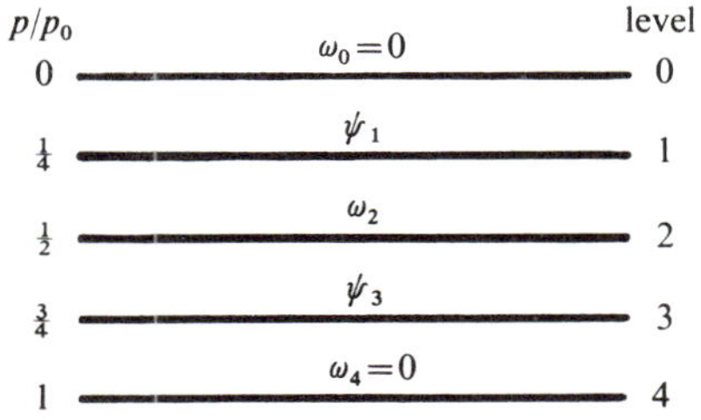

If ω_2 is eliminated between (11.10) and (11.11), two equations for ψ_1 and ψ_3 result which may be integrated numerically.

11.3   Primitive equation models

The baroclinic model described in § 11.2 although less restrictive than the barotropic model of §11.1 still involves a number of approximations. In particular the motion is assumed to be quasi-geostrophic. This restriction can be removed by writing the equations in a more basic form, namely: the horizontal momentum equations (7.8)

$$\frac{\partial \mathbf{V}}{\partial t} + \mathbf{V} \cdot \nabla V + \omega \frac{\partial \mathbf{V}}{\partial p} + f \mathbf{k} \wedge \mathbf{V} = -\nabla \Phi + \mathbf{F} \tag{11.12}$$

the continuity equation (7.20)

$$\mathbf{V} \cdot \mathbf{V} + \frac{\partial \omega}{\partial p} = 0 \tag{11.13}$$

the hydrostatic equation (1.2)

$$\frac{\partial \Phi}{\partial p} + \frac{1}{\rho} = 0 \tag{11.14}$$

and the thermodynamic equation (11.6)

$$\frac{\partial}{\partial t}\left(-\frac{\partial \Phi}{\partial p}\right) + \mathbf{V} \cdot \mathbf{V}\left(-\frac{\partial \Phi}{\partial p}\right) - \frac{B\omega}{g\rho^2} = \frac{1}{c_p\rho}\frac{dS}{dt} \tag{11.15}$$

In these equations $\mathbf{V}$ is the horizontal wind and $\mathbf{V}$ refers to differentiation at constant pressure.

The first equations (11.12) to (11.15) ((11.12) is of course two equations) involve five unknowns, two components of $\mathbf{V}$, ω, Φ and ρ.

Although ordinary sound waves have been eliminated from the solutions of this set of equations by the use of the hydrostatic equation (11.14) and the neglect of vertical accelerations, solutions remain for a wide range of different motions including for instance the acoustic and the gravity waves described in § 8.3. These solutions are eliminated for the simplified models of § 11.1 and § 11.2 through the quasi-geostrophic approximation.

With primitive equation models care has to be taken that the solutions are not swamped by spurious gravity or acoustic waves which may arise from errors in the initial data or from computational instability. To overcome these problems it is necessary to introduce an initial motion field which satisfies (11.12); actual observations of wind must not be employed as they stand. Further a time step has to be chosen for the integration process which is short enough to avoid computational instability for the fast-moving gravity or acoustic waves. For this reason time steps need to be considerably shorter than with quasi-geostrophic models with similar horizontal resolution. Primitive equation models, therefore, demand much more computer time than comparable quasi-geostrophic ones.

In (11.12) a frictional term **F** has been included and in (11.15) the diabatic heating term. A number of physical processes are involved in these terms. Further an adequate description of the boundary conditions at the surface is required. Some indications of how these matters may be dealt with in a numerical model are presented in the sections which follow. More detail may be found in the works listed in the bibliography.

11.4   Inclusion of orography

It is necessary in a model to take into account the varying height of the surface. In the isobaric co-ordinate system where pressure is the vertical co-ordinate, the varying surface pressure due to the varying height of the surface is not easy to handle. For numerical models, therefore, a co-ordinate σ is employed, defined as

$$\sigma(x, y, t) = \frac{p(x, y, t)}{p_s(x, y, t)} \tag{11.16}$$

where p_s is the surface pressure. This system is known as the *sigma co-ordinate system* (fig. 11.2, problem 11.1). σ is a non-dimensional parameter; the lower boundary is always at $\sigma = 1$ where the vertical σ velocity $\dot{\sigma} = 0$.

11.5   Convection

As integrations in the model proceed situations occur where the vertical temperature gradient is super adiabatic. These are especially likely in regions where there is strong heating of the surface. The simplest way to make allowance for the effect of convection is to carry out a process of *convective adjustment*. The temperature profile is adjusted to the dry adiabatic (§ 1.4) under conditions of low water vapour content, and to the saturated adiabatic (§ 3.2) when the relative humidity is equal to or close to 100%. This adjustment has to be made, of course, in such a way that the total energy is unchanged, i.e. the sum of the total potential energy and latent heat released by the adjustment must be zero.

In a more elaborate scheme of convective parameterization (see Dickinson and Temperton, 1984) the assumption is made that within each grid box where there is vertical instability an ensemble of buoyant convective plumes with varying characteristics (temperature, humidity, cross-sectional area) exists; these entrain air from the environment and ascend until no longer buoyant. If condensation occurs, the latent heat released can increase the buoyancy of the plumes; clouds possessing different vertical extents result from the ensemble. Regions of descent compensate for the ascending plumes. As in the simpler convective scheme adjustments are made to ensure the total energy is unchanged in the process.

11.6   Moist processes

Water vapour evaporated from the surface and moved about by atmospheric motions plays a very significant part in the atmosphere's energy budget because of (1) the release of latent heat on condensation and (2) the formation of clouds which alter the radiation budget (cf. chapter 6). A numerical model, therefore, needs to keep track of the water content in vapour and liquid form. An equation rather similar to the continuity equation may be

Fig. 11.2. The σ-co-ordinate system for the 15-level Meteorological Office Model.

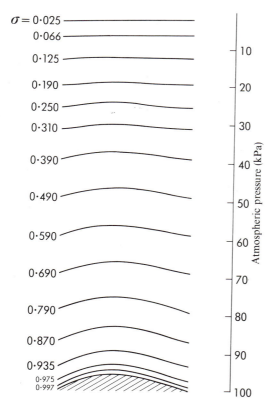

written for the water vapour mixing ratio m

$$\frac{\partial}{\partial t}(\rho m) + \mathbf{V} \cdot \mathbf{V}(\rho m) + \omega \frac{\partial(\rho m)}{\partial p} = -C + E + D \tag{11.17}$$

where C is the rate of condensation, E the rate of evaporation into the volume and D is a diffusion term (cf. § 11.8). The rate of condensation C is determined by comparing the water vapour mixing ratio m with the saturation mixing ratio m_s appropriate to the atmospheric temperature. If $m > m_s$, the excess water is assumed to condense. Precipitation is therefore formed which can fall towards the surface; if it falls through unsaturated air, allowance is made for some of the precipitation to evaporate.

Associated with the rate of condensation C there is a latent heat release LC, where L is the latent heat of condensation; if further, temperatures are below 0°C the ice phase is involved and the latent heat of melting also needs to be taken into account. These latent heat terms contribute to the term dS/dt in the energy equation (11.15).

11.7 Radiation transfer

The radiative processes which contribute to atmospheric heating and which need to be included in the adiabatic term in (11.15) divide into two parts (§ 2.1) namely the absorption of solar radiation and the exchange of terrestrial radiation. Simple examples of calculations for both of these were given in chapter 4.

Absorption of solar radiation in a clear atmosphere is mainly by ozone, water vapour and carbon dioxide. Detailed calculations of the amount of absorption by an atmospheric path for any of these gases may be made using the methods outlined in chapter 4 and the data in appendices 8, 9 and 10. For incorporation into numerical models, however, these methods are too expensive in computer time and further simplification is necessary. The various bands are fairly well separated from each other so that for the sun at zenith angle θ the solar energy absorbed in an atmospheric path is

$$\cos \theta \{ f_1[u(O_3)] + f_2[u^*(H_2O)] + f_3[u^*(CO_2)] \} \tag{11.18}$$

where f_1, f_2 and f_3 are empirical functions. The path length

$$u = \sec \theta \int_{\text{path}} c\rho \, dz \tag{11.19}$$

(c being the absorber concentration at height z where the density is ρ) is employed in the calculation of ozone absorption without any correction (cf. appendix 9). For water and carbon dioxide, however, allowance needs to be made for the dependence of absorption on the pressure p (§4.4); the simplest way of doing this is to employ a modified path length

$$u^* = \sec\theta \int_{\text{path}} c\rho\left(\frac{p}{p_0}\right)^x dz \quad (0.5 < x < 1) \tag{11.20}$$

which is scaled so as to be appropriate to a standard pressure p_0. A suitable value of x in the range $0.5 < x < 1$ is chosen empirically (fig. 11.3).

Fig. 11.3. (a) The absorption of solar radiation by an atmosphere containing water vapour as a function of modified path length u^* (11.20) in g cm^{-2} with $x = 0.675$ for a range of values of p/p_0, showing that the pressure-scaling employed is a good approximation.

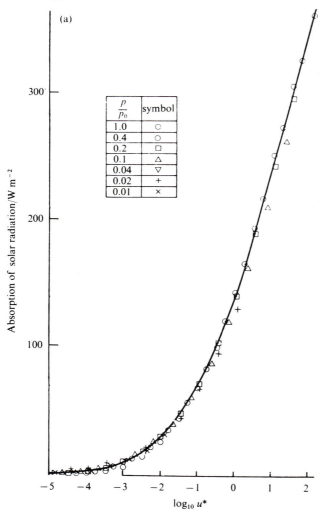

(b) The same as (a) but for an atmosphere containing carbon dioxide with modified path length u^* (in g cm^{-2}) and with $x = 0.68$.

(c) The emissivity of a slab of atmosphere at 220 K containing water vapour showing the usefulness of the modified path length u^* ($x = 0.7$) in g cm^{-2} as a parameter. The wavenumber range 550–800 cm^{-1} has been omitted since in that range CO_2 absorption is dominant. (All after Manabe & Wetherald, 1967)

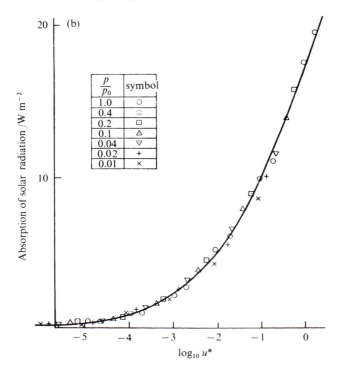

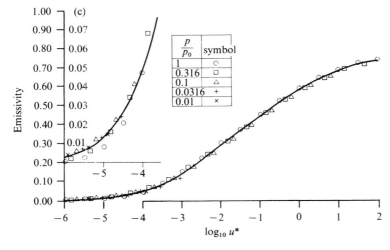

Absorption of solar radiation not only occurs in the direct beam but also in the diffuse beam reflected or transmitted by clouds or reflected from the surface. The effect of absorption of the upward travelling radiation is calculated by employing (11.18) for the total path and including a factor to allow for the average direction of the diffuse beam.

The transfer calculation for terrestrial radiation also needs to be enormously simplified. In chapter 4 band models to assist in the integration over frequency were described. Even these are, however, too time consuming for normal use with large numerical models. A simpler approach is to divide the spectrum into a smaller number of spectral intervals denoted by Δv_i and following (4.23) to write for the upward or downward flux at level z

$$F(z) = \sum_i \left\{ \int_{z'=z}^{z'=z_0} \pi B_i(z') \, d\tau_i[u^*(z', z_0)] + \pi B_i(z_0)\tau_i[u^*(z, z_0)] \right\} \qquad (11.21)$$

where $B_i(z')$ is the Planck function appropriate to the temperature at level z' integrated over Δv_i, u^* is the pressure-scaled path length as defined by (11.20) (with $\sec \theta = \frac{5}{3}$, cf. §4.7) for the absorbing gas (either CO_2, H_2O or O_3) appropriate to the spectral interval Δv_i. τ is an empirical transmission function determined by experiment and appropriate to a reasonable range of atmospheric conditions (fig. 11.3) and z_0 denotes the boundary which may be the ground or the surface of a cloud. From the upward and downward fluxes so calculated, the radiative contribution to the heating term dS/dt in (11.15) may be found by differencing the fluxes as in (2.4).

It will be clear from the discussion of chapter 4 that (11.21) involves a lot of approximation both in its basic formulations (see for instance Rodgers, 1967) and in its use of a single parameter u^* to describe the absorbing path for each gas and spectral interval. However, for calculations in the lower atmosphere the accuracy achieved by using (11.21) even with only one or two spectral intervals is satisfactory, especially considering the much larger errors which may arise from the difficulty of adequately specifying the atmospheric parameters in the models.

11.8   Inclusion of clouds

The presence of clouds exerts a profound influence on both the short-wave and the long-wave radiation fields. The cloud field, however, is often very complex having structure on all scales including the sub grid scale of the model. Great simplification is therefore necessary in the model description of clouds and their effect on the radiation fields. In a typical scheme (Slingo (ed.), 1985) clouds are specified by fractional cover and by height of upper and lower surfaces (assumed to coincide with model levels). All clouds apart from those at high level are given an albedo of 0.6; high level clouds are assumed to have an albedo of 0.2 (cf. §§ 6.3 and 6.4). They are also assumed to absorb a fraction of 0.1 (0.05 for high clouds) and to transmit 0.3 (0.75 for high clouds) of incident solar radiation. Their upper and lower surfaces are taken to be 'black' absorbers and emitters of long-wave radiation apart from high clouds which are given an emissivity of 0.75. Allowance is made in the scheme for multiple reflections between different cloud layers or between cloud layers and the surface (problem 11.4).

An even bigger difficulty than that of adequately simplifying the cloud description is to arrange for clouds to be prescribed from the model variables of humidity, temperature, pressure, etc. in a realistic manner. Because of this difficulty many models have avoided the problem altogether by specifying within the model average cloud amounts drawn from climatological records and dependent only on height, latitude and season. With such a scheme, however, no feedback between cloud amount, radiation and other processes can, of course, occur. For very shortrange forecasts (up to 1 or 2 days) this feedback is probably not important, but for longer range forecasts or for climatological studies it is essential for the cloud specification to depend on the model variables (cf. § 13.3). For convective cloud this can be done as part of the convective scheme by allowing the cloud amount and extent to depend on the amount of condensation. For other cloud types, empirical relationships can be set up between the cloud amount and the humidity field; the coefficients in the relationship being 'tuned' by comparing the cloud climatology generated by the model with the observed cloud climatology (Slingo, 1980).

11.9 Sub grid scale processes

We saw in § 9.6 that there is interaction in the atmosphere between motions on a wide range of scales. With a numerical model, therefore, which inevitably possesses a finite grid size it is necessary to make allowance for the effect of motions on smaller scales than the grid size. The terms $\mathbf{F}$ in (11.12) and D in (11.17) were included to allow for such transfer of momentum and water vapour by the small scale eddies; also there will be a contribution to the term dS/dt in (11.15) from the eddy transfer of heat. The simplest formulation of both horizontal and vertical eddy transfer employs eddy diffusion coefficients K which when multiplied by the appropriate gradients give the transfer, e.g. (9.6). Typical values of K for the troposphere for sub grid scale parameterization lie in the range 1–10 $m^2 s^{-1}$. The Ks can, of course, be allowed to vary with height, the static stability, or with other quantities describing the local flow. In particular, K for momentum can be formulated in a way which preserves the k^{-3} spectral distribution of turbulent energy (§ 9.6).

11.10 Transfer across the surface

Momentum, heat and water vapour are transferred across the earth's surface; equations for the transfer of each are required.

First some method is required to arrive at the surface temperature. Since the effective heat capacity of a land surface is small, its temperature T_s is determined by its local heat balance, i.e.

$$(1-A)I_S + F^\downarrow - \sigma T_s^4 + Q_s - LE = 0 \tag{11.22}$$

where σ is the Stefan–Boltzmann constant, A is the surface albedo, I_S the incident solar radiation (after allowing for absorption or reflection in the atmosphere or the clouds above as described in § 11.7), $F^\downarrow$ the downward flux of terrestrial radiation from the atmosphere or from clouds calculated as described in § 11.7, Q_s the vertical transfer of heat from the atmosphere to the surface by eddy processes, E the rate of evaporation of water vapour from the surface and L the appropriate latent heat. The emissivity of the surface for the terrestrial radiation is assumed to be unity. Methods of calculating Q_s and E are described below. For the ocean surface, by contrast, an infinite heat capacity is assumed and the temperature of the sea surface is assumed to be constant with time; values for it are included in the data initially given to the model.

If the model's lowest level is close to the surface and well within the boundary layer, say at 100 m altitude or so, the simplest way of describing the flux of momentum across the surface or the surface stress τ is (following Manabe, 1969 and Holloway & Manabe, 1971)

$$\tau = -\rho c_D |\mathbf{V}| \mathbf{V} \tag{11.23}$$

where ρ is the density, c_D the drag coefficient (see problem 11.3), $\mathbf{V}$ the vector wind appropriate to the first level. Similar expressions may be written for the flux of heat Q_s and water vapour E, namely

$$Q_s = c_p \rho c_D |\mathbf{V}| \Delta\theta \tag{11.24}$$

and

$$E = \rho c_D |\mathbf{V}| \Delta m \tag{11.25}$$

where $\Delta\theta$ and Δm are respectively the difference in potential temperature and water vapour mixing ratio between the surface (assumed saturated at surface temperature) and the first level. Equations (11.23) to (11.25) are applicable to neutral conditions; the effect of different stabilities can be taken into account by allowing c_D to vary with stability. Equation (11.25) also only applies to a wet surface. This is, of course, all right over the ocean. Over the land allowance must be made for variation in the soil moisture content W (mass per unit area) which can be found from the equation

$$\frac{\partial W}{\partial t} = P - E \tag{11.26}$$

where P is the precipitation rate determined from the moist processes of § 11.6 and E the evaporation rate. If W exceeds a given value the excess water is assumed to run off. If W exceeds a certain lower value W_k the surface is assumed wet; if $W < W_k$ the evaporation rate is assumed to be reduced by the ratio W/W_k from the value given by (11.25). Snow cover can be allowed for in a similar way to soil moisture.

The simple formulation of equation (11.23) applies reasonably well over the sea or level terrain. To estimate the momentum exchange due to flow over mountains, it is necessary to model the generation of gravity waves (§ 8.3 and problem 11.5), their propagation upwards, depending on the vertical stability, and the subsequent absorption of their momentum in the upper troposphere and lower stratosphere where the waves break. Palmer *et al.* (1986) have set up a description of the whole process suitable for incorporation into a numerical model.

11.11 Forecasting models

As an example of the models currently employed at a major forecasting centre, the models in use at the Meteorological Office, Bracknell, will be briefly described. (For further information see Dickinson & Temperton, 1984 and Gadd, 1985.) The *Global Model* is a grid point primitive equation model with 15 levels in the vertical (fig. 11.2) and with horizontal spacing of 1.5° in latitude ($\Delta\phi$) and 1.875° in longitude ($\Delta\lambda$). At the corners of the grid (fig. 11.4) the surface

Fig. 11.4. The arrangement of variables on the model's horizontal grid. For the global model of the Meteorological Office $\Delta\phi = 1.5°$ and $\Delta\lambda = 1.875°$. (After Gadd, 1985)

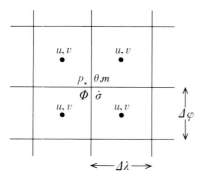

pressure (p_*), potential temperature (θ), humidity mixing ratio (m) are predicted while the geopotential (Φ) and vertical velocity ($\dot{\sigma}$) may be diagnosed. The horizontal velocity components (u, v) are predicted at the centre of the grid boxes. A 15-minute time step is employed for the integration. At high latitudes where the grid spacing in longitude becomes small, a filtering process is introduced to ensure computational stability.

Observations are assimilated into the model (Atkins & Woodage, 1985) gradually during a run of the model for the six hours preceding the analysis time. This run starts from the forecast situation predicted for the appropriate time from the model's previous run. The process is illustrated in fig. 11.5. It ensures that useful information contained in the previous model run is not thrown away and it also ensures, by the gradual introduction of new observations, that the process is computationally stable. An example of a forecasting product from the global model is shown in fig. 11.6.

Despite the increase in the speed of computers (fig. 11.7), large forecasting models are computer limited. A 24-hour forecast with the global model involves about 10^{11} numerical operations and takes 4 minutes on the Cyber 205 computer.

The Meteorological Office also runs a *limited area model* with half the horizontal resolution, i.e. about 75 km instead of about 150 km for the global

Fig. 11.5. Schematic diagram of data assimilation. The ordinate ψ represents the state of the model or of the atmosphere. The abscissa is the model time in hours. T is the time at which observations are to be assimilated and $T-6$ is the start of the assimilation process from the model prediction for time $T-6$ based on the previous forecast run. ψ^{observed} is the state of the atmosphere as determined from observations at time T; ψ^{model} is the model prediction for time T based on the previous forecast run. The line ending $\psi^{\text{model}+\text{data}}$ illustrates the assimilation process. At each time step (steps are separated by Δt) the model is modified so as to move in small increments towards the observations. (After Atkins & Woodage, 1985)

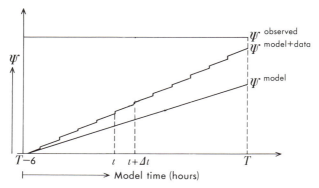

model. The formulation is otherwise identical as for the global model from which boundary conditions for the limited area model are obtained. An example of a forecast with the model is shown in fig. 11.8.

A further model for operational forecasting at the Meteorological Office is the *mesoscale model* – a model with a 15 km horizontal grid extending over the immediate regions of the British Isles. The model possesses 16 levels, 5 in the lowest kilometre so as to give a good representation of the boundary layer. An important difference from larger scale models is that hydrostatic balance is not imposed so vertical accelerations are possible. The model is particularly aimed at assisting in short-term and local forecasting; further details are given in Golding (1984).

11.12 Other models

In the models described so far in this chapter, the model parameters

Fig. 11.6. A chart showing upper winds (at 39 000 ft) and temperatures issued for the use of aviation directly from the global numerical forecast model at the Meteorological Office. Temperatures in $-°C$: the wind at 30N, 40°W is 65 kt from the NW.

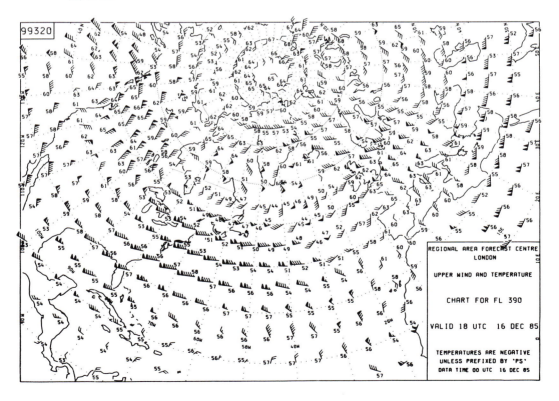

have been specified at grid points and the integrations carried out by finite difference methods applied to the array of grid points. Another type of model which has some advantages and which is widely employed is the *spectral model* in which horizontal variations are represented in terms of a series of spherical harmonics. Such models are particularly valuable for studies of the upper atmosphere. Here most of the variations are described by the first few terms of such a series, i.e. those describing the first few wave numbers (the wave number refers to the number of complete cycles of a variation which occurs around a latitude circle). A substantial amount of computing time can therefore be saved by the use of such transformations.

Also valuable for particular studies are *two-dimensional models*, i.e. models in which atmospheric parameters are averaged around latitude circles so that only variations with height and latitude are considered. The main difficulty with such models is that of devising means for dealing with the transport of heat and momentum by the large scale eddies which have to be related in simple ways to zonally averaged quantities. Two dimensional models are particularly useful in upper atmospheric modelling in which changes in composition due to complex photochemical reactions (cf. § 5.5) have to be

Fig. 11.7. The growth in speed in number of floating-point operations per second of computers employed by the Meteorological Office for numerical models of the atmosphere. A straight line is also drawn with a growth of a factor of 10 every 4.5 years.

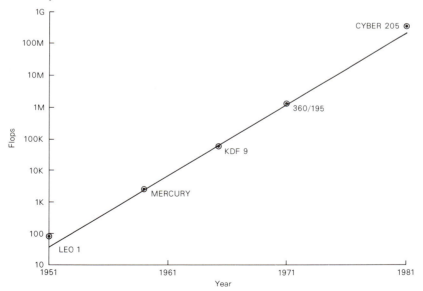

included and for which integrations over long periods (e.g. several years) are required (see e.g. Harwood & Pyle, 1975).

The models we have described in this chapter have shown a great deal of success in describing the dynamics of the atmosphere and in operational forecasting. Nevertheless, these models contain some fundamentally unsatisfactory features. For instance in § 11.3 it is pointed out that the numerical solutions derived from the set of primitive equations not only describe the development of the relatively slowly varying *synoptic* field which is the primary interest of the weather forecaster, but they also describe a much wider class of solutions including gravity waves. Means of damping out these unwanted solutions have to be applied, a process which becomes increasingly difficult as the horizontal resolution of the model is increased. This higher resolution is needed as was pointed out in § 11.11 in order to include explicitly as much as possible of the interaction between the synoptic flow and the smaller scale detail and to attempt to describe more adequately the almost discontinuous structures (e.g. fronts, inversions and the tropopause) which are, in effect, part of the synoptic pattern. However, the highest resolutions currently realized in models do not match the scale of these structures.

Current models distinguish between resolved and unresolved features purely on the basis of horizontal scale and time scale. A procedure which would overcome this difficulty would be to employ a *Lagrangian*† description of the air parcels allowing their properties to evolve slowly, but without restricting the spatial scale. It has been shown by Cullen, Norbury and Purser (1986) that an effective means of describing the evolution is to employ an energy principle which assumes that the air parcels are arranged at all times close to a minimum energy configuration. The configuration evolves slowly in time under the action of the forcing fields. The method has so far been successfully applied to the modelling of fronts (Cullen & Purser, 1984). Further details are given in problems 11.7 and 11.8.

Problems

11.1 Show that the following equation will transform from the isobaric vertical co-ordinate system to the sigma system

$$\mathbf{V}_p = \mathbf{V}_\sigma - \frac{\sigma}{p_s} \nabla p_s \frac{\partial}{\partial \sigma} \tag{11.27}$$

where $\mathbf{V}$ refers to horizontal differentiation only. Hence derive the horizontal momentum equation

$$\frac{dV}{dt} + f\mathbf{k} \wedge \mathbf{V} = -\nabla\Phi + \frac{\sigma}{p_s} \nabla p_s \frac{\partial\Phi}{\partial\sigma} \tag{11.28}$$

Fig. 11.8. Illustrating a 30-h forecast with the limited area model of the Meteorological Office, (a) Surface pressure for 0001Z 10 October 1985, (b) Analysis for 0600Z 11 October 1985,

(a)

(b)

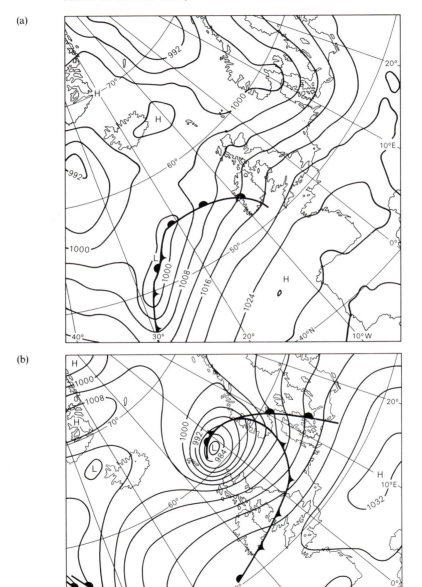

and the continuity equation

$$\mathbf{V} \cdot (p_s \mathbf{V}) + p_s \frac{\partial \dot{\sigma}}{\partial \sigma} + \frac{\partial p_s}{\partial t} = 0 \qquad (11.29)$$

where $\quad \dot{\sigma} = \dfrac{d\sigma}{dt}$

11.2 Consider a vertical column of unit cross-section above a location where the surface pressure is p_s. By considering the convergence of mass into the column show that

$$\frac{\partial p_s}{\partial t} = - \int_0^1 \mathbf{V} \cdot (p_s \mathbf{V}) \, d\sigma$$

† A *Lagrangian* description is one which follows the flow of the fluid as opposed to an *Eulerian* description which describes the flow referred to a co-ordinate system fixed in space.

Fig. 11.8. (c) Forecast starting from (a) for 0600Z 11 October 1985. Note the accurate prediction of the deepening of the depression and its movement over the period.

(c)

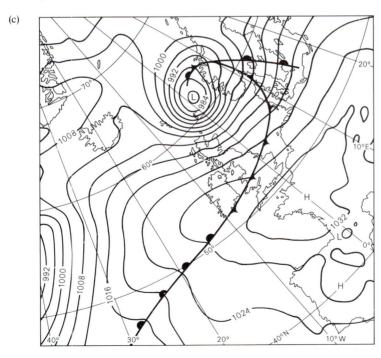

Show that the same result may be obtained by integrating the continuity equation (11.29).

11.3 Show from (9.15) that under neutral conditions c_D in (11.23) is given by

$$c_D = \left[\frac{\kappa}{\ln(z/z_0)} \right]^2 \tag{11.30}$$

Hence find its value if $z = 100$ m, $z_0 = 1$ cm and $\kappa = 0.4$.

11.4 A uniform cloud of reflectivity R is present above a surface with albedo α. Taking into account multiple reflections between the cloud and the surface and ignoring absorption or variations of reflectivity with angle show that if the proportion of the incident solar flux S absorbed by the surface is written as $(1 - \alpha_e)S$ where α_e is an effective reflectivity, then

$$\alpha_e = \frac{\alpha(1 - R)}{(1 - \alpha R)} \tag{11.31}$$

11.5 The generation of gravity waves by mountains is an important mechanism leading to exchange of momentum between the atmosphere and the surface. If waves are generated with horizontal and vertical velocitues u and w (both varying as $\exp i(\omega t + kx + mz)$) the stress τ_s at the surface will be given by equation 9.6

$$\tau_s = \overline{\rho u w}$$

Show that the vertical velocity w generated by a mountain of height h will be proportional to $k\bar{u}h$. Then employing equation 8.24 and the result of problem 8.21, show that providing $m \gg \frac{1}{2}H$ and $m \gg k$

$$\tau_s = \alpha k \rho \bar{u} \omega_B h^2 \tag{11.32}$$

where α is a constant which may be chosen empirically.

In the formulation of gravity wave drag for incorporation into the Meteorological Office numerical model Palmer *et al.* (1986) have given αk the value 2.5×10^{-5} m^{-1} and limited the value of $\overline{h^2}$ to $(400 \text{ m})^2$. Consider some real orography and discuss the reasonableness of these values.

11.6 A height-like variable z can be defined such that it is related to the pressure p by the expression

$$z = \frac{H_0}{\kappa}\left(1 - \left(\frac{p}{p_0}\right)^\kappa\right) \tag{11.33}$$

where $H_0 = RT_0/Mg$ is the scale height under surface conditions where $p = p_0$ and the temperature $T = T_0$ and where $\kappa = (\gamma - 1)/\gamma$ (cf. § 3.1).

Show from the hydrostatic equation and equation (3.4) that

increments in z are related to increments in geometrical height h by the expression

$$\theta \, dz = \theta_0 \, dh \tag{11.34}$$

where θ is the potential temperature and $\theta_0 (= T_0)$ is the potential temperature at the surface. Hence show that for an adiabatic atmosphere $z = h$. Show also that the maximum value of z is about 28 km. What is the percentage difference between z and h for an isothermal atmosphere at the level where $z = H_0$?

11.7 For an incompressible, hydrostatic, atmosphere under the Boussinesq assumption (i.e. the atmosphere is assumed to be of uniform density apart from buoyancy effects) and with the vertical coordinate z defined as in problem 11.6, derive the following set of equations.

$$\frac{dM}{dt} = -\frac{\partial \phi}{\partial y} \tag{11.35}$$

$$\frac{dN}{dt} = \frac{\partial \phi}{\partial x} \tag{11.36}$$

$$\frac{\partial \phi}{\partial z} = \frac{g\theta}{\theta_0} \tag{11.37}$$

$$\frac{d\theta}{dt} = 0 \tag{11.38}$$

$$\frac{\partial u}{\partial x} + \frac{\partial v}{\partial y} + \frac{\partial w}{\partial z} = 0 \tag{11.39}$$

Here ϕ is the geopotential, θ the potential temperature and

$$M = v + fx$$
$$\text{and} \quad N = -u + fy \tag{11.40}$$

are components of what is called the *absolute momentum* (Hoskins, 1975). They are given that name because, taking into account the rotation of the earth, in the absence of forcing terms (e.g. pressure gradients), M and N are the appropriate conserved quantities which replace the velocities v and u. Appropriate boundary conditions to go with the set of equations (11.35)–(11.39) are $w = 0$ at $z = 0$ and H.

   In deriving an expression for the energy, it is useful to remember that for a fluid of nearly uniform density given by $\rho(h)$ at level h, the difference between its potential energy and that which would occur if the fluid were of completely uniform density is given by

$$\int \frac{g(\rho(z) - \rho_0)h}{\rho_0} \rho_0 \, dV$$

Hence show that for a Boussinesq atmosphere under the assumptions being made here, an appropriate expression for the sum E of the kinetic and potential energies is:

$$E = \int_V \left[\frac{1}{2}(u^2 + v^2) - \frac{g\theta z}{\theta_0} \right] \rho_0 \, dV \tag{11.41}$$

where the integral is over the whole domain of interest. By substituting for M and N show that E can be written

$$E = \int \left[\frac{1}{2}f^2(y^2 + x^2) + \frac{1}{2}(M^2 + N^2) - Nfy - Mfx - \frac{g\theta z}{\theta_0} \right] \rho_0 \, dV \tag{11.42}$$

The procedure which is being developed by Cullen *et al.* (1986) is to determine the evolution of the flow from a balanced initial state by minimizing the energy integral (11.42) in a way which is consistent with the equations (11.35)–(11.39). The volume of an air parcel in x, y, z space is conserved by (11.39), the potential temperature θ of the parcel is conserved by (11.38). No fluid is allowed to cross the boundaries. The evolution of the flow is driven by the pressure gradient terms in (11.35) and (11.36) and the fluid responds by rearrranging itself in such a way as to minimize the energy.

The simplest case to consider is one-dimensional in which there is no horizontal motion and $\theta = \theta(z)$ only. The energy is then potential energy only. Show by exchanging an air parcel near the surface with one at a higher level that the potential energy term in (11.42) is minimized by moving to an arrangement in which θ increases monotonically with z. This is similar to the process described in § 3.5 when considering available potential energy.

Now going back to equation (11.42) we note that the first term in the energy integral is independent of the flow and the second does not change under rearrangement because as we mentioned above, in the absence of forcing M and N are conserved. It is therefore required to minimize the remaining terms, i.e.

$$\int \left(-Nfy - Mfx - \frac{g\theta z}{\theta_0} \right) \rho_0 \, dV \tag{11.43}$$

By extension of the one-dimensional argument, it can be shown that, for (11.43) to be minimized under all possible rearrangements it is necessary for N to increase monotonically with y, M with x and θ with z.

An important theorem proved by Cullen and Purser (1984) is that there exists a unique minimizing arrangement for a fluid divided up into finite elements in which the quantities M, N and θ in (11.43) must be related to a parameter η such that

$$fM = \frac{\partial \eta}{\partial x} \qquad fN = \frac{\partial \eta}{\partial y}, \qquad \frac{g\theta}{\theta_0} = \frac{\partial \eta}{\partial z} \qquad\qquad (11.44)$$

Show that if

$$\eta = \phi + \tfrac{1}{2}f^2(x^2 + y^2) \qquad\qquad (11.45)$$

then M and N are as defined in equation (11.40) but with u and v replaced by their geostrophic values.

With M and ∂N defined in this way, in an arrangement with finite elements in each of which the values of M, N and θ are constant, show that the slopes of the boundary between two elements whose values of M, N and θ differ by ΔM, ΔN and $\Delta\theta$ respectively, are given by

$$\frac{dx}{dz} = \frac{g}{f\theta_0} \frac{\Delta\theta}{\Delta M} \qquad\qquad (11.41)$$

Fig. 11.9. Illustrating a very simple model of sea breeze front.

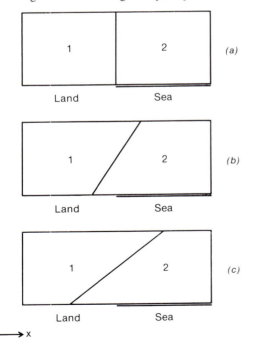

$$\frac{dy}{dz} = \frac{g}{f\theta_0} \frac{\Delta\theta}{\Delta N}$$
(11.42)

Note that equations (11.41) and (11.42) are essentially the thermal wind equations (cf. § 7.6)).

11.8   To illustrate the application of the methods described in problem 11.7, consider an atmosphere divided into two finite elements of equal volume (fig. 11.9) initially at rest, possessing identical values of θ and possessing values of M differing only because the mean value of x differs between the two elements. Initially, the lower boundary of one element is land the other sea. As the land heats up during the day, show that the boundary between the two elements (i.e. the sea breeze front) changes in the way shown in fig. 11.9 until an equilibrium state is reached (fig. 11.9(c)) in which the two elements contain equal areas of the land surface.

12

Global observation

12.1   What observations are required?

To provide adequate and accurate initial conditions for numerical models and to enable comparison to be made between the models and the real atmosphere, detailed knowledge of the atmospheric state over the whole atmosphere is required continuously in time. We need to know the three dimensional fields of motion, of density and of composition. Clearly, measurement in complete detail is not possible. Further, some parameters are related to each other. For instance, away from the equatorial regions, the geostrophic approximation is a good one so that motion need not necessarily be measured independently if the pressure field is specified. Also, if the field of temperature is known together with the pressure at some reference surface, the density and pressure can be deduced at all levels from the hydrostatic equation (1.4) (problem 12.1). The following specification of what is required is considered realistic from the measurement point of view; it is a shortened version of the specification prepared for the first Global Atmospheric Research Programme global experiment (chapter 13).

Atmospheric state parameters	Accuracy (RMS error)
Wind components	$\pm 3 \, \text{m s}^{-1}$
Temperature	$\pm 1 \, \text{K}$
Pressure of reference level	$\pm 0.3\%$
Water vapour pressure	$\pm 0.1 \, \text{kPa}$
Sea surface temperature	$\pm 0.2 \, \text{K}$

Measurements of these parameters (or their deduction from other observations) are required every 12 hours, with at least one measurement every 100 km in the horizontal and at least eight data levels in the vertical (surface, 90, 70, 50, 20, 10, 5 and 2 kPa). Information is also required on precipitation, cloud cover, surface conditions (e.g. snow or ice cover) and elements of the radiation budget.

12.2   Conventional observations

Over the land areas of the world much of the information about the atmosphere comes from a high density of surface observations of pressure, temperature, humidity, wind and cloud cover together with measurements twice per day from a large number of *radiosondes*. These are balloon-borne packages containing simple instruments (fig. 12.1) for the measurement of pressure, temperature and humidity, together with a radio transmitter and a radar reflector so that the package can be tracked to give details of the wind. Measurements up to $\sim 30\,\text{km}$ ($\sim 1\,\text{kPa}$) are possible with the radiosonde. A

Fig. 12.1. Compatibility between observations from different meteorological instruments is of paramount importance as observations from many different sources provide the input for the global analyses from which global forecasting models begin their integrations. Illustrated here are five radiosondes (which measure pressure, temperature, humidity, wind) used by different countries attached to the same balloon so that direct comparison between their measurements can be made. (Courtesy of Meteorological Office)

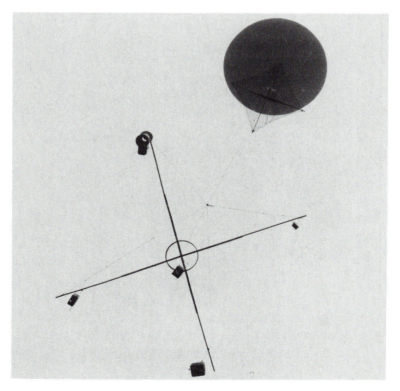

much more sparse network of rocket-sondes – sondes released from high altitude (up to ~90 km) rockets which then descend on parachutes – enables some observations to be made at higher levels.

More specialized measurements, especially of atmospheric composition, are possible from larger balloons, aircraft or rockets. Spectroscopic observations, particularly of absorbed sunlight, from the surface or from various altitudes also provide information regarding composition. The study of meteor trails and other techniques has enabled density and wind in the 70–100 km region to be deduced.

Radar is an important technique for the observation of precipitation and its distribution (fig. 12.2). A typical radar wavelength for this purpose is

Fig. 12.2. Rainfall display obtained from the UK weather radar network at 0900 GMT on 6 August 1981, as displayed in real time on a television monitor. Large scale ascent due to baroclinic instability and small scale rising motion due to buoyant convection are both important in the generation of rain, but there is also a high degree of organization associated with dynamical processes occurring on an intermediate scale, the so-called mesoscale. The radar echo patterns obtainable from networks of ground based radars is well suited to portraying these mesoscale features.
The rainfall patterns, normally depicted in colour, are shown here as three shades of grey, corresponding to light (< 1 mm h^{-1}), moderate to heavy (1–16 mm h^{-1}) and very heavy (> 16 mm h^{-1}) rain. Coastlines are shown electronically as white dots and the outer limit of radar coverage is represented by circles centred about four radar sites in England. (From Browning (1982))

3 cm. Water drops and ice crystals scatter back radar energy, the scattered intensity being strongly dependent on droplet size (to the sixth power of the diameter) and on whether the precipitation particles are of water or ice. Calibration factors, preferably measured directly by comparing some radar signals with actual rainfall rates, need to be applied to convert a map of radar echoes to one of rates of precipitation.

12.3   Remote sounding from satellites

The great advantage of a satellite as a measurement platform is that good coverage in space and time can be obtained. From a *geostationary* satellite, orbiting at 35 000 km altitude so that it remains directly above a fixed point on the equator, continuous observation of about a quarter of the atmosphere is possible. A *polar orbiting* satellite such as Nimbus or Tiros in a circular orbit at about 1000 km altitude makes about 14 orbits a day and can view all parts of the atmosphere at least twice per day.

Radiation from the earth–atmosphere system reaches an orbiting satellite over a wide range of wavelengths. In the ultraviolet, visible and near infrared, solar radiation is scattered and reflected from the surface, from clouds, from aerosol (particles suspended in the atmosphere) and from molecules. In the infrared and microwave regions, at wavelengths almost completely separated from those where solar radiation is present (fig. 2.1), radiation is emitted again from the surface, clouds and molecules. Over this wide range of wavelengths a great deal of information is contained about the structure and composition of the atmosphere below. Interpretation of these *remote sounding* observations is often complex and difficult, but as we have seen, they possess the enormous advantage compared with conventional observations that a satellite can cover a very large area in a short time.

The first weather satellite was launched in 1960; it carried television cameras for viewing clouds. For the first time complete pictures of the cloud associated with large weather systems were seen. Such information is now produced routinely from a large number of satellites. Detailed cloud pictures of Mars (fig. 10.15), Venus (fig. 2.7) and Jupiter (fig. 13.6) from space probes have also provided a surprisingly large amount of information about the circulation of their atmospheres.

Infrared imaging systems on satellites enable the emitted radiation field to be mapped in various spectral intervals. In atmospheric *window* regions, for instance between 10 and 12 μm in wavelength, the radiance received corresponds quite closely to the Planck black-body function at the surface or cloud-top temperature (fig. 12.3). Suitable small corrections have to be made for atmospheric transparency and surface emissivity (problem 12.2). Measurements over broad spectral regions provide information about the earth's radiation budget over different areas (e.g. fig. 4.4).

By combining measurements at different wavelengths at which the properties of the surface or of clouds differ slightly, more detailed information can be retrieved. Fig. 12.4 illustrates the identification at night of the detailed distribution of fog over England and Wales by such a method.

Measurements at higher spectral resolution in different infrared regions give more precise information about the earth's temperature structure and composition. These we shall consider in turn in the following sections.

12.4 Remote sounding of atmospheric temperature

At any frequency in the infrared where an atmospheric constituent possesses strong absorption the radiation intensity leaving the top of the

Fig. 12.3. Image of the earth in the infrared (10–12 μm) taken by NOAA satellite in geostationary orbit on 5 July 1975. Warm surfaces are dark and cold surfaces bright. Notice the hurricane in the centre of the picture.

11:15 05JL75 32A-Z 0006-1640 FULL DISC IR

atmosphere is a function of the distribution of the emitting gas and the distribution of temperature through the atmosphere.

Constituents such as carbon dioxide with strong absorption bands at 15 μm and 4.3 μm and molecular oxygen with absorption near 5 mm wavelength in the microwave region, are very nearly uniform mixed, at least up to levels ~ 90 km altitude. Provided that local thermodynamic equilibrium (LTE) applies (§ 5.6) the emitted intensity in these bands can be considered to be dependent only on the atmospheric temperature distribution.

Fig. 12.4. This image of the southern UK and the near continent taken at 0433 GMT on 23 October 1983 was constructed from two infra-red channels of the Advanced Very-High Resolution Radiometer aboard NOAA-7. The differences between the brightness temperatures viewed at 3.7 μm and 10.8 μm have been coded as shades of grey.

Objects with the same brightness temperature in the two channels such as the land and the ocean surfaces are shown as mid-grey. The brightness temperatures of thin cirrus cloud and certain small but hot sources such as gas flares are higher at 3.7 μm than at 10.8 μm. They are depicted as dark shades. The emissivity, and hence the brightness temperature, of water clouds is lower at 3.7 μm than at 10.8 μm. These appear as light shades. Low lying water cloud or fog can therefore readily be distinguished from the surface below (after Eyre *et al.* (1984)). The marked lines at the coasts are artifacts caused by a slight mis-registration of the two channels. (Courtesy of the Meteorological Office)

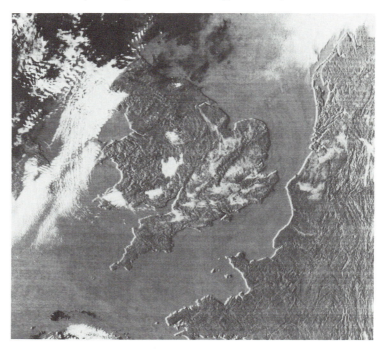

At a given wavenumber $\tilde{\nu}$ the intensity of radiation $I_{\tilde{\nu}}$ (known as the *radiance*) received by a satellite-mounted radiometer viewing vertically downwards is given by (4.21)

$$I_{\tilde{\nu}} = \int_0^\infty B_{\tilde{\nu}}(T) \frac{d\tau_{\tilde{\nu}}(z, \infty)}{dz} \, dz + B_{\tilde{\nu}}(T_s)\tau_{\tilde{\nu}}(0, \infty) \tag{12.1}$$

where T_s is the surface temperature.

The variable $y = -\ln p$, where p is the pressure in atmospheres, is a more convenient height-dependent variable to use, so that (12.1) can be written

$$I_{\tilde{\nu}} = \int_0^\infty B_{\tilde{\nu}}(T)\kappa(y) \, dy + B_{\tilde{\nu}}(T_s)\tau_{\tilde{\nu}}(0, \infty) \tag{12.2}$$

$\kappa(y)$ is known as the *weighting function*. For a spectral region with a uniform

Fig. 12.5. 'Weighting functions' appropriate to a radiometer sounding the atmosphere by observing radiation emitted vertically upwards from the atmosphere for (a) an atmosphere with uniform absorption coefficient, problem 12.3, (b) a frequency in the wing of a pressure broadened spectral line, problem 12.4, (c) an Elsasser band, problem 12.5. Note the greater vertical resolution of (b). p_m is the pressure at which the functions peak.

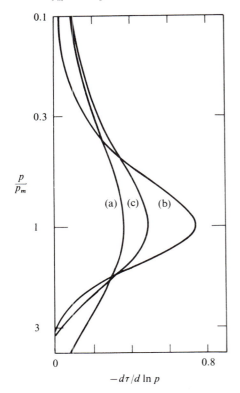

absorption coefficient $k_{\tilde{v}}$ independent of altitude (problem 12.3 and fig. 12.5),

$$\kappa(y) = k_{\tilde{v}} c p g^{-1} \exp\left(-k_{\tilde{v}} c p g^{-1}\right) \tag{12.3}$$

where c is the mass mixing ratio of the absorbing constituent. The function $\kappa(y)$ has a peak at $k_{\tilde{v}} c p g^{-1} = 1$, i.e. where the optical depth measured from the top of the atmosphere is unity.

Inspection of fig. 12.6 illustrates how remote temperature sounding is achieved. Moving from the atmospheric window near $800 \, \text{cm}^{-1}$ (12.5 μm) to smaller wavenumbers, the mean absorption coefficient of carbon dioxide gradually increases so that the average level being monitored moves up in

Fig. 12.6. Thermal emission from the earth plus atmosphere emitted vertically upwards and measured by the infrared interferometer spectrometer on Nimbus 4, (a) over Sahara, (b) over Mediterranean, (c) over Antarctica. The radiances of black bodies at various temperatures are superimposed. (From Hanel *et al.*, 1971)

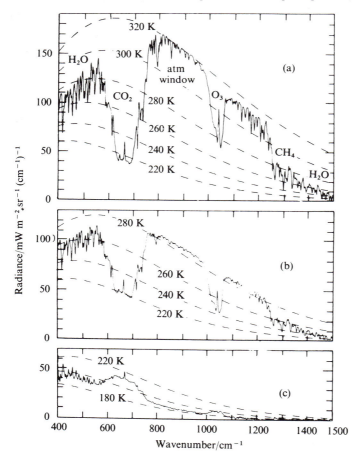

altitude until the most absorbing region is reached – the Q branch near 667 cm^{-1} – where the radiation largely originates in the stratosphere. For figs. 12.6(*a*) and (*b*), the temperature decreases with altitude to the tropopause, then increases again, while for fig. 12.6(*c*) the temperature increases with altitude all the way up. In fig. 12.7 the application of results to the remote sounding of tropospheric temperature is illustrated. Notice also that the spectra of fig. 12.6 contains information from which the water vapour and ozone distributions may be inferred. Similar observations have also been made of the atmosphere of Mars for which, being almost pure carbon dioxide, measurements in the 15 μm carbon dioxide band are particularly appropriate.

It is important that radiance measurements be made with adequate accuracy. For the 15 μm region, to obtain temperatures to 1 K, radiance must be measured with a maximum error of considerably less than 1% (problem 12.6). This implies careful radiometric calibration and also means, for conventional instrumentation, that the spectral resolving power cannot be very high. It will be noticed in fig. 12.6 that individual rotational lines cannot be resolved so that much information, particularly about the high atmosphere, is

Fig. 12.7. Temperature profile (full line) retrieved from measurments of radiance in the 4.3 μm and 15 μm CO$_2$ bands by the high resolution infrared sounder and from measurements in the 5 mm O$_2$ band by the scanning microwave spectrometer on Nimbus 6.

The sounding is 48°N, 15.2°E on 23 August 1975; the circles are a nearby radiosonde. The presence of 3% of high cloud and 24% of medium cloud was deduced from the retrieval process. (After Smith, 1976)

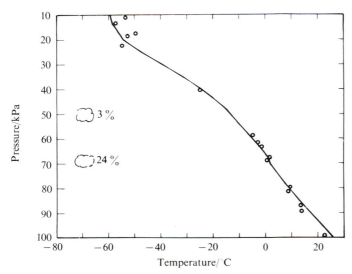

being lost. Houghton & Smith (1970) have overcome this problem by employing a less conventional spectroscopic technique – selective chopping with absorbing gas cells – achieving a spectral resolution approximately equal to that of the absorption lines themselves. Fig. 12.8 shows the weighting functions appropriate to the selective chopper radiometer on Nimbus 4; fig. 12.9 and figs. 8.5 and 8.6 show results from that instrument illustrating the global coverage obtained for an altitude region of the atmosphere largely inaccessible to conventional observation. A further development of the technique known as the pressure modulator radiometer enables remote sounding temperature measurements to be made to ∼90 km altitude.

The major problem in the interpretation of radiance measurements in infrared regions is the very variable presence of clouds. Considerable effort has gone into developing techniques for retrieving temperature profile information when broken cloud is present. Because non-precipitating clouds are largely transparent at millimetre wavelengths, measurements of oxygen emission at 5 mm wavelength do not suffer from this drawback, and are, therefore, particularly important for monitoring the troposphere (cf. fig. 12.7).

Fig. 12.8. Weighting functions for the six channels of the selective chopper radiometer on Nimbus 4. (After Abel *et al.*, 1970)

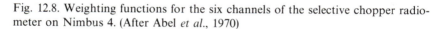
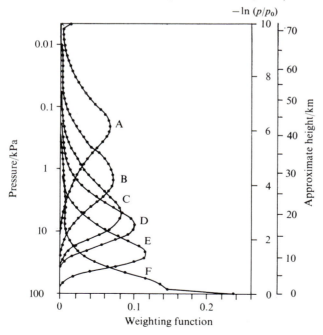

12.5   Remote measurements of composition

Various absorption bands of water vapour are available for observation in the infrared and microwave regions; given temperature information from carbon dioxide or oxygen emission, some details of the water vapour distribution may be inferred by similar techniques. The ozone distribution may also be studied in this way.

Also important for remote sounding observations of ozone, particularly at high levels, is the ultraviolet region. Solar radiation back-scattered from the atmosphere is strongly affected by ozone absorption in the 200 to 300 nm region. To illustrate the method consider a simplified situation in which an ultraviolet spectrometer is observing the radiation leaving the atmosphere in a vertical direction, the sun being overhead (fig. 12.10). At the levels considered attenuation of solar radiation by ozone absorption will be much greater than that due to scattering (problem 12.11); further, single scattering only will be considered.

At a given frequency v the intensity of solar radiation at the top of the

Fig. 12.9. Temperature (K) cross-section of the atmosphere from 80N to 80S deduced from radiance measurements from the selective chopper radiometer on Nimbus 5 and the pressure modulator radiometer on Nimbus 6 for 4 August 1975.

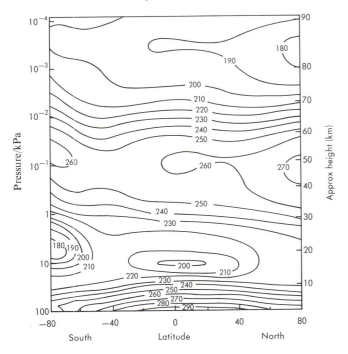

atmosphere is $I_{S\nu}(0)$ and at a level where the pressure is p will be

$$I_{S\nu}(p) = I_{S\nu}(0) \exp \{ -k_\nu n(p) \} \tag{12.4}$$

k_ν being the absorption coefficient per molecule of ozone and $n(p)$ being the number of molecules above the level of pressure p.

From the molecules in a layer of thickness dp at level p the radiation scattered vertically upwards will be proportional to $I_{S\nu}(p)$ and to dp, say it is $aI_{S\nu}(p)\,dp$ (fig. 12.10). Further attenuation will occur on traversing the return path to the satellite, so that neglecting any scattering from the lower atmosphere, the measured intensity will be

$$I_\nu = I_{S\nu}(0)a \int_0^{p_0} \exp \{ -2k_\nu n(p) \}\, dp \tag{12.5}$$

For purposes of illustration consider a uniform mixing ratio of ozone, i.e. $n(p) = n'p$ say, n' being a constant and

$$I_\nu = I_{S\nu}(0)a \int_0^\infty p \exp(-2k_\nu n'p)\, d(\ln p) \tag{12.6}$$

The quantity under the integral is very similar to the weighting function of (12.3); it has a peak at $p = (2k_\nu n')^{-1}$. By choosing a set of wavelengths for which k_ν is different, it therefore becomes possible to monitor the ozone concentration at different levels. Such measurements have been made from the back-scatter ultraviolet spectrometer mounted on Nimbus 4 (Krueger, Heath & Mateer, 1973).

To make remote sounding measurements of the distribution of minor constituents such as CH_4, N_2O, CO, NO, NO_2, HNO_3, etc., which are involved

Fig. 12.10

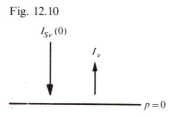

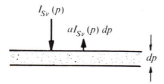

in ozone photochemistry (§ 5.5) or pollution studies, it is necessary to observe radiation emitted by them (or absorbed from the incident solar radiation) from the atmospheric limb (figs. 12.11 and 12.12 and problems 12.9 and 12.10). A long atmospheric path is thereby available for observation which is helpful when looking for the effect of constituents which may be present only in a few parts in 10^9. The limb view also avoids the problem of the very varying background of the earth's surface or of clouds.

On the Upper Atmosphere Research Satellite (UARS) (fig. 12.12) whose purpose is to observe details of stratospheric structure and composition, the instruments between them cover a wide spectral range; most of them are directed to view the atmospheric limb.

In addition to measurements of gaseous composition it is also important to observe the distribution of liquid water in the atmosphere and the distribution of precipitation. Because of the varying spectral properties of water vapour and liquid water in the microwave part of the spectrum, observations near 1.5 cm wavelength can help to provide such information, especially over the oceans (Staelin *et al.* 1976, Wilheit *et al.* 1977).

12.6   Other remote sounding observations

Because of the breakdown of the geostrophic approximation in tropical regions, good measurements of the wind field are required, at least at low latitudes. Observation of the motion of suitable clouds using images from geostationary satellites provides good information at two levels in the

Fig. 12.11. Illustrating limb sounding of the earth's atmosphere. Measurements of emission from the atmosphere's limb have the advantages of (1) a very long emitting path is viewed so that constituents present in very small concentrations can be studied, (2) near-zero radiation background beyond the limb.

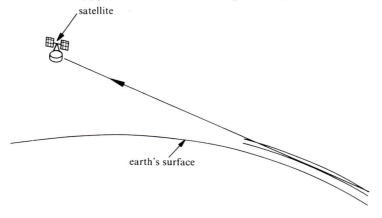

troposphere, namely near 90 kPa where small convective clouds can be tracked and near 20 kPa where cirrus clouds occur. Such observations are, of course, limited to regions where suitable clouds are present; observation in other regions can be made by following the motion of free floating balloons or of sondes dropped from aircraft.

The observations of the state of the land or sea surface is also important to provide for general circulation models appropriate information regarding the lower boundary especially information relevant to the exchange of momentum, heat and water-vapour across that boundary. Accurate measurements of surface temperature especially over the oceans requires careful radiometry at infrared wavelengths (problem 12.2). Imagery at microwave wavelengths can easily be interpreted to distinguish between water and ice (Webster *et al.*, 1975) and also provide information about surface wetness and vegetation cover.

Fig. 12.12. The Upper Atmospheric Research Satellite (UARS) carries a number of remote sounding instruments for measuring atmospheric structure and composition namely the Cryogenic Limb Array Etalon Spectrometer (CLAES), the Improved Stratospheric and Mesospheric Sounder (ISAMS), the Microwave Limb Sounder (MLS), the Wind Imaging Interferometer (WINDII), the Halogen Occultation Experiment (HALOE) and the High Resolution Doppler Imager (HRDI); the HALOE and the HRDI cannot be seen from this view. Other instruments measure solar radiation and solar particles. The Multimission Spacecraft (MMS), the Tracking and Data Relay (TDRS) antenna are also shown. UARS is a large satellite about 9 m long and weighing about 5 tons.

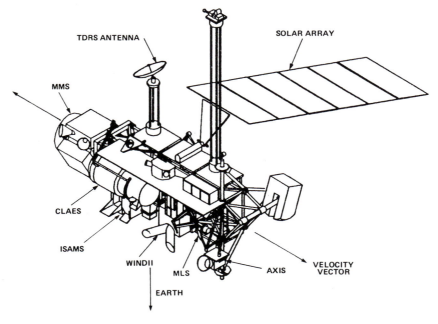

Conditions at the surface of the earth can also be observed from the character of the back-scattered signals from satellite-borne radars. The speed and direction of the surface wind over the ocean can be measured by a scatterometer observing the radar returns, at a wavelength of a few cm, from capillary waves on the sea surface.

Remote sounding from instruments carried on space probes is also an important means for investigating the structure of planetary atmospheres. The structure and circulation of the atmospheres of Venus (cf. problem 12.14 and figs. 12.16 and 7.10), Mars and Jupiter have been investigated in this way. Further details of remote sounding instrumentation and observations from satellites and space probes can be found in Houghton, Taylor & Rodgers (1984).

12.7 Observations from remote platforms

To deduce the atmospheric density field, in addition to global measurements of temperature, the pressure at a reference surface is required. Conventional surface pressure observations are only very sparse over the oceans.

Free floating buoys in the oceans with appropriate instrumentation can be interrogated from orbiting satellites which act as a communication link. Satellite links can also arrange for automatic communication with other remote observation platforms on land or with commercial aircraft.

12.8 Achieving global coverage

The deployment of polar orbiting satellites and five geostationary satellites required by operational meteorologists to provide adequate coverage of the globe from space is illustrated in fig. 12.13. The polar orbiter satellites in the Tiros-N series (fig. 12.14) of the National Oceanic and Atmospheric Administration carry a number of instruments for imaging at different wavelengths and for the remote observation of atmospheric temperature and water vapour. The data available each day from these satellites together with conventional observations add up to a large amount (fig. 12.15); their assimilation into numerical models is a complex process that requires to take into account the particular characteristics (including the error characteristics) of each type of measurement. This global system of observation known as World Weather Watch is co-ordinated internationally and is being continually improved as the quality of observations improves and as other techniques for observation are developed.

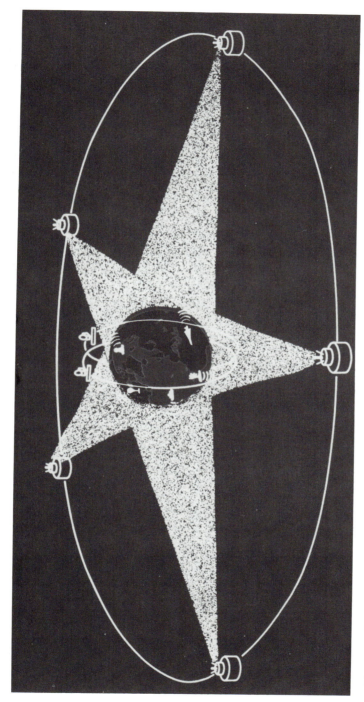

Fig. 12.13. Illustrating the satellite observing system for World Weather Watch; five geostationary satellites and two or more polar orbiting satellites. Remote sounding observations are made from the satellites; they are also employed as communication links to relay observations from ships, buoys, aircraft and remote observation stations.

Problems

12.1 The density near the 10 kPa level is deduced from surface pressure measurement combined with temperature measurements throughout the rest of the atmosphere. What is the percentage error in density resulting from (a) a 1 K error in temperature at 10 kPa, (b) a 1 K error in temperature at all levels below 10 kPa, (c) a 3% error in pressure measurement at the surface?

12.2 From the information contained in problem 4.17, and assuming the water vapour mixing ratio at any altitude is proportional to p^3 where p is the atmospheric pressure in atm, calculate the transmission of a vertical column of atmosphere in the 11 μm window.

   If $B(p)$ is the black-body function at 11 μm appropriate to the temperature at the level where the pressure is p atm, for an atmospheric lapse rate, $B(p)$ is approximately equal to $B_0 p$ where B_0 is the black-body function at the surface temperature. Using this expression and (4.21) calculate the intensity leaving the top of the atmosphere at 11 μm.

Fig. 12.14. The Tiros-N Spacecraft. The body of the satellite is about 3.71 m long.

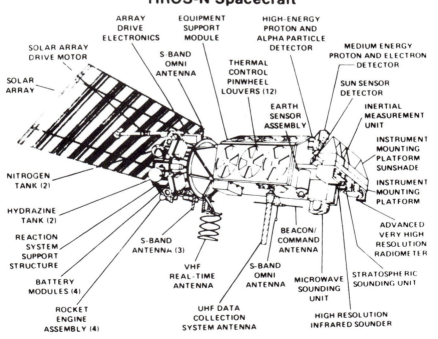

TIROS-N Spacecraft

Fig. 12.15. Distribution of data as received at the European Centre for Medium Range Weather Forecasts for 4 June 1979, 12 GMT ± 3 h. The diagrams show, respectively, the coverage of satellite winds from tracking cloud images (top two on left), of surface observations (bottom on left), of satellite temperature soundings (top right), of pilot balloons and radiosonde ascents, of information from buoys interrogated from satellites (middle two on right) and of observations from aircraft (bottom on right). (After Bengtsson *et al.*, 1982)

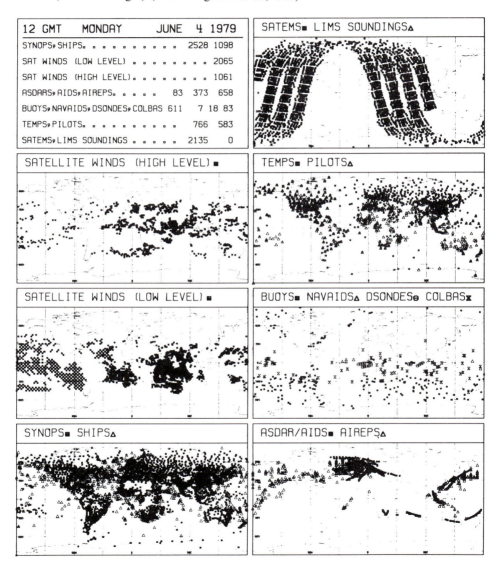

Hence estimate the error in the surface temperature deduced from a satellite observation if atmospheric absorption is ignored.

12.3 Obtain the expression in (12.3) for the 'weighting function' for an atmosphere with uniform absorption coefficient $k_{\tilde{\nu}}$.

12.4 Show that in the wing of a collision broadened line of strength s, and width γ, the absorption coefficient is given by (cf. (4.5))

$$k_{\tilde{\nu}} = \frac{s\gamma}{\pi(\nu - \nu_0)^2}$$

Hence show that the weighting function $\kappa(y)$ appropriate to such a frequency is of the form

$$\kappa(y) = 2\left(\frac{p}{p_m}\right)^2 \exp\left[-\left(\frac{p}{p_m}\right)^2\right]$$

where $\quad p_m^2 = \dfrac{2g\pi p_0(\nu - \nu_0)^2}{s\gamma_0 c}$

(quantities as defined in (4.7) and (12.3)).

12.5 A useful model of a set of absorption lines is that due to Elsasser which assumes uniform line strength and spacing. For this model for collision broadened lines the transmission of a path between the level at pressure p and the top of the atmosphere is given by

$$\tau = 1 - \frac{2}{\pi^{1/2}} \int_0^{\beta p} \exp(-x^2)\, dx$$

where β depends on line strengths, widths, spacing and absorber concentration. Show that, for this case, the weighting function is given by

$$\kappa(y) = \left(\frac{2}{\pi}\right)^{1/2} \frac{p}{p_m} \exp\left(-\frac{p^2}{2p_m^2}\right)$$

12.6 Calculate from the Planck function for a source at 240 K the percentage accuracy in radiance measurements which is required to achieve a 1 K accuracy in temperature for wavelengths of (a) 4.3 μm, (b) 15 μm, (c) 5 mm.

12.7 A radiometer observing the atmosphere possesses an optical system with a collecting area A, field of view Ω steradians, spectral bandwidth $\Delta\tilde{\nu}$ centred at wavenumber $\tilde{\nu}$, and mean transmission τ_0. Incident radiation from a black body of radiance $B_{\tilde{\nu}}$ is chopped sinusoidally.

Show that the signal:noise ratio for measurement in t seconds is

$$\frac{S}{N} = \frac{B_{\tilde{\nu}} A \Omega \Delta \tilde{\nu} \tau_0 t^{1/2}}{2^{3/2}(\text{NEP})}$$

where NEP is the noise equivalent power of the detector (i.e. the r.m.s. radiation power incident on the detector which gives a signal equal to the noise in a 1 Hz bandwidth).

12.8 The noise equivalent temperature (NET) of a radiometer is the change of source temperature which causes a change in signal just equal to the noise. Calculate for a source temperature of 240 K the NET for an instrument with the following characteristics:

$A = 100 \text{ cm}^2$, $\Omega 10^{-4} \text{ sr}$, $\tau_0 = 0.5$,

$\Delta \tilde{\nu} = 1 \text{ cm}^{-1}$ at 667 cm^{-1}, $\text{NEP} = 10^{-10} \text{ W}$,

$t = 4 \text{ s}$.

12.9 A satellite mounted radiometer is observing the limb of the atmosphere. Show that the path of atmosphere traversed through the limb (i.e. $\int \rho \, dx$ where x is along the line of sight) is approximately 70 times that in a vertical path above the tangent point of the path.

12.10 Show that for the path of problem 12.9 the mean pressure along the path as defined by (4.15) is $2^{-1/2} p_0$ where p_0 is the highest pressure along the path. What is the horizontal distance between points where the pressure $2^{-1/2} p_0$?

12.11 From the formula in problem 6.14, for a path near 60 km altitude and $0.27 \mu\text{m}$ wavelength compare attenuation due to Rayleigh scattering with that due to ozone absorption (ozone concentration at 60 km, 10^{10} cm^{-3}, ozone molecular absorption cross-section 10^{-17} cm^2, other information in the appendices).

12.12 In an atmosphere with spherical symmetry where density ρ and hence refractive index n varies with distance r from the centre of the sphere, for a ray passing through the atmosphere the quantity nd is a constant where d is the length of the perpendicular from the centre of the sphere to the tangent to the path at the point where the refractive index is n. From this expression determine the actual position of the lowest point of a ray passing through the atmosphere to have (a) 10 km, (b) 20 km, as its lowest point.

12.13 A spacecraft emitting radio signals is occulted by the planet Venus. By studying the time taken for reception of the signals it is possible to trace the paths of rays through the Venus atmosphere during occultation and so to deduce the variation of density with altitude. Find the highest pressure on the Venus atmosphere from which such rays will emerge. (Refractive index of CO_2 at STP for radio frequencies $= 1.00049$.)

12.14 Suppose that the selective chopper radiometer as built for Nimbus 4 (weighting functions for the earth's atmosphere in fig. 12.8 covering the range from 0.1 to 100 kPa) were employed to view the atmosphere of Venus. From equation 12.3 and data in table 1.1 calculate the range of pressure within the Venus atmosphere which could be observed. In fig. 12.16 vertical temperature profiles for the earth and for Venus are compared.

Fig. 12.16. Vertical temperatures profiles of temperature versus pressure at 30N latitude as measured on Venus by the VORTEX experiment and on earth by the Nimbus 7 Stratospheric and Mesospheric sounder. Further details of both instruments in Houghton, Taylor & Rodgers 1984.

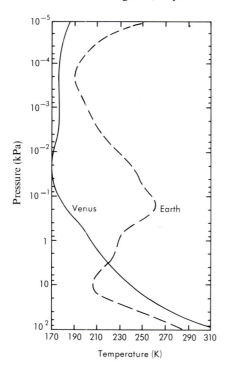

13

Atmospheric predictability and climatic change

13.1　Short-term predictability

Numerical models, as described in chapter 11, can with some confidence predict the atmospheric state for periods ahead up to a few days. For longer periods ahead, detailed prediction is difficult for two main reasons, firstly, the problem of describing the initial fields with adequate precision and, secondly, the deficiencies which are inherent in the models themselves. The most fundamental of the latter is the parameterization of sub grid scale motions. Even when, as larger computers become available, the grid size is reduced, there will always be smaller scale motion to be considered which, because of the lack of a comprehensive theory describing the interactions between different scales which always occur in turbulent motion, can only be dealt with in a crude empirical manner. The existence of this fundamental difficulty raises the question as to how predictable in principle is the atmosphere's general circulation.

To attack this problem the Global Atmospheric Research Programme (GARP) was formulated, culminating in 1979 in the Global Weather Experiment, the purpose of which was, over a period of a year, to observe the atmosphere as completely as possible so that the performance of numerical models could be tested (fig. 13.1) and light thrown upon the fundamental question of predictability. The Global Weather Experiment is the largest internationally co-ordinated scientific experiment man has yet undertaken. Some idea of the large quantities of data acquired during the Experiment will be apparent from fig. 12.15. Analysis of these data is still being carried out. It is clear, however, that improvement of the quality and coverage of the initial data, improvement of the description within the models of the physical processes, and improvement of the numerical methods, all lead to an improvement in prediction by the numerical models (fig. 13.2). Amongst meteorologists there is an expectation in the future of useful forecasts being provided of the detailed structure of the atmosphere's general circulation out to between 7 and 14 days ahead.

Fig. 13.1. The global experiment of the Global Atmospheric Research Programme (GARP).

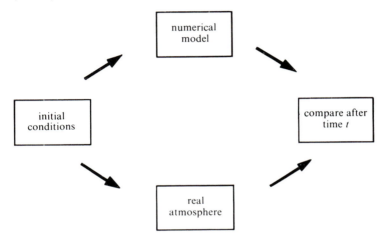

Fig. 13.2. RMS error of the field of geopotential height at 50 kPa as forecast for 48 hours ahead by the models at the U.K. Meteorological Office.

    Mean values over 12 months (September to August) are shown for different models: 3-level model (◆), 10-level model (○) 15-level model. (●). Also shown are the errors for 24-hour forecasts in 1966 and for the 72-hour forecast in 1983.

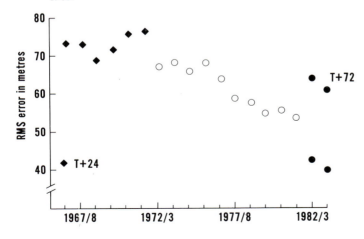

13.2   Variations of climate

Climate is loosely defined as averaged weather; more specifically the character of given climate regimes are defined through the statistical properties of appropriate parameters – averaged values and deviations therefrom over the region and timescale of interest.

Fig. 13.3. Trends in global climate: (a) Climate of the last half-million years deduced from measurements of oxygen isotope ratio in plankton shells which relate to global ice volume (after Hays, Imbrie & Shackleton, 1976). (b) Climate of last 1000 years estimated from evidence relating to east European winters (after Lamb, 1966). (c) Climate of last 100 years as evidenced by changes in average annual temperature of Northern hemisphere (after Mitchell, 1977).

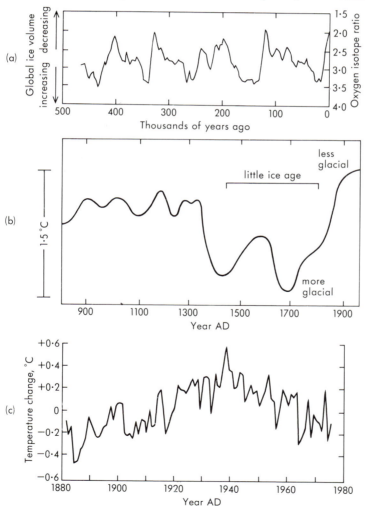

Substantial changes in the atmosphere's circulation have taken place in historic and prehistoric times (fig. 13.3). The large number of possible causes for such changes are illustrated in fig. 13.4. These causes can be extra-terrestrial, i.e. changes in solar radiation, or they can arise from couplings within the atmosphere itself or between the atmosphere, the ocean, the ice or the land surface. Because of the long periods associated with the deep ocean circulation, or changes in the physical or chemical properties of the surface, the characteristic times associated with different possible mechanisms can be anything from a few years to periods comparable with the age of the earth itself.

13.3   Atmospheric feedback processes

Processes exist in the atmosphere possessing characteristics of both positive and negative feedback. As an example of positive feedback consider a land mass which is covered by ice or snow. Because of the resulting high albedo of the surface much of the incident solar radiation is reflected back out of the atmosphere, thus (in the absence of any other mechanisms) reducing the surface temperature, leading to possible further increases in ice or snow cover. For an example of negative feedback consider an increase in solar radiation incident on the earth's surface which leads to a higher surface temperature and hence to increased evaporation. Resulting from the higher water content of the atmosphere is an increase in cloudiness which in turn reduces the solar radiation reaching the surface.

Fig. 13.4. Schematic illustration of the components of the coupled atmosphere–ocean–ice–earth climatic system. (From GARP Publication Series No. 16 (see Bibliography))

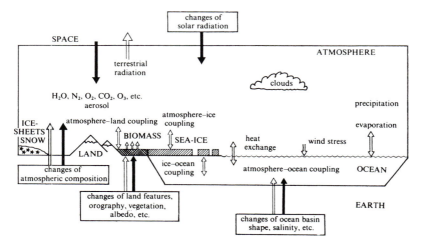

A more complicated example of a feedback process is *cloud–radiation feedback*, which can be either positive or negative. The presence of a cloud can affect the radiation field in two ways. Because clouds reflect solar radiation back to space (chapter 6) the amount of solar radiant energy at the surface is reduced by the presence of cloud. On the other hand, clouds, because they contain liquid water or ice, are strong absorbers and emitters of infrared radiation. Clouds, therefore, have a blanketing effect on long-wave radiation emitted by the earth's surface and tend to decrease the net loss of long-wave radiation by the surface.

Fig. 13.5 and problem 13.6 shows that for high clouds which have low liquid water or ice content and a relatively low albedo, the blanketing effect dominates tending to lead to higher surface temperatures. For low clouds, on the other hand, with high liquid water content and high albedo, the loss of

Fig. 13.5. Equilibrium surface temperature distributions in K calculated from an atmospheric radiative transfer model as a function of cloud amount for three cloud layers and for July conditions at 35N. The clouds are assumed to possess liquid water amounts in a vertical path of 140, 140 and $20 \, \mathrm{g \, m^{-2}}$ and to occupy layers 91–85, 63–55 and 38–30 kPa respectively. (After Stephens & Webster, 1981)

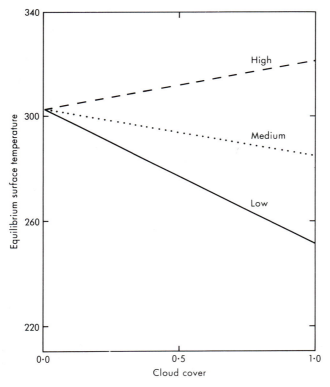

energy through the reflection of solar radiation dominates leading to lower surface temperatures. Predicting the height and type of cloud is therefore critical if the sign of the cloud–radiation feedback is to be correctly predicted.

Even more complex feedback processes arise from interactions between the atmosphere and the land surface (for instance, its vegetative cover) and between the atmosphere and the ocean. The atmosphere–ocean interaction is particularly important because the ocean covers over 70% of the earth's surface and directly absorbs more solar energy than any other component of the climate system – it absorbs twice as much as the atmosphere, for instance. Further, the ocean circulation is largely driven by forcing from the atmosphere through the action of surface winds. In turn the atmosphere is very sensitive to the exchange of heat energy from the ocean over its lower boundary, either in terms of sensible heat or latent heat (through evaporation). Of even more importance is the input of water vapour to the atmosphere from the ocean surface, bringing with it the potential for latent heat release through condensation elsewhere in the atmosphere.

The sea surface temperature is a sensitive indicator of these inputs from the ocean to the atmosphere. There is now a substantial amount of evidence, theoretical and empirical, that the atmosphere's circulation is influenced by sea surface temperature anomalies, especially those that occur in tropical regions. The best known of these anomalies is the El Nino phenomenon which occurs every few years off the coast of S. America and which is strongly correlated with climate anomalies in many parts of the world. Through the World Climate Research Programme (WCRP), a programme following from the Global Atmospheric Research Programme, a concentration of research effort is being directed towards better descriptions and better understanding of these interactions.

Because of the existence of these complicated feedbacks in the atmosphere questions have been raised as to how stable is the atmosphere's average state and whether there are other quasi-stable states between which the atmosphere may fluctuate. Inspection of the record of climatic history suggests that such may in fact be the case.

13.4 Different kinds of predictability

In a discussion of weather forecasting or the prediction of climate change it is useful to distinguish between various kinds of predictability. Lorenz (1975) classified two types of predictability. For predictability of the *first kind*, the prediction is dependent on the initial conditions as well as on the boundary conditions. For predictability of the *second kind* the prediction is dependent on the boundary conditions only.

For a prediction model based on a physical system containing the atmosphere only, the boundary conditions are the solar constant and the properties of the surface (e.g. the sea surface temperature, the albedo and the soil moisture). For more complex models aimed at climate prediction, the parts of the system under consideration will include the ocean, the ice and vegetation. If each component of the system and the interactions between the components are properly modelled, conditions at the interfaces between the components (e.g. the sea surface temperature, the surface albedo, the soil moisture, the sea ice extent) are determined by the model itself. In this case, the appropriate boundary conditions are the solar constant, the size, mass and rotation of the planet, some properties of the land surface (e.g. orography, land use, etc.) and the mass and composition (except water vapour content) of the atmosphere.

For weather forecasting of large-scale features up to, say, 7–14 days ahead an atmospheric general circulation model is employed with fixed boundary conditions. A prediction of the detailed behaviour of the atmosphere's circulation is looked for. This is prediction of the first kind. For slightly further ahead, say up to about a month, the boundary conditions are still essentially fixed. Although prediction of the precise detail of the circulation is not possible on this timescale some skill exists for prediction of some statistical properties of the circulation which are still dependent on the initial conditions. This is also, therefore, prediction of the first kind.

Turning to longer time scales, memory of the initial conditions is lost and the statistical properties (or the climate) of the prediction depend only on the boundary conditions. Climate predictions with such models are predictions of the second kind.

Of particular importance is the prediction of the sensitivity of climate to changes which might result from human activities. For instance, the burning of fossil fuels leads to an increase in the carbon dioxide content of the atmosphere which increases the blanketing by the atmosphere of long-wave radiation from the surface and which therefore in turn tends to lead to increased surface temperatures. Increases in the concentration of other minor constituents can have a similar effect (problem 13.7), as can changes in aerosol content (problem 13.8).

13.5 Jupiter's Great Red Spot

The Great Red Spot on Jupiter has been in existence at least over the 300 years since it was first observed by Robert Hooke in 1664. It is a large circulation about 15 000 km in diameter. Other similar circulations, though not so large, are also evident in Jupiter's atmosphere (fig. 13.6). Strong dynamical similarities exist between these features in the atmosphere of Jupiter and stable, closed baroclinic eddies which can be produced in laboratory experiments on thermal convection in a rotating fluid subject to internal heating or cooling (fig.

Fig. 13.6. (a) A mosaic of Voyager 1 images of Jupiter projected to show the planet as seen from below the South Pole. The Great Red Spot appears in the top right of the image. Other long-lived anticyclonic eddies appear as trains of white oval spots on lines of constant latitude. (b) and (c) Streak photographs of the circulation in a rotating laboratory annulus subject to internal heating and sidewall cooling showing regular baroclinic eddies. The value of Φ (problem 13.9) is 0.73 for the conditions in which the isolated eddy shown in (b) is formed. For (c) $\Phi = 0.54$. (After Read & Hide, 1984)

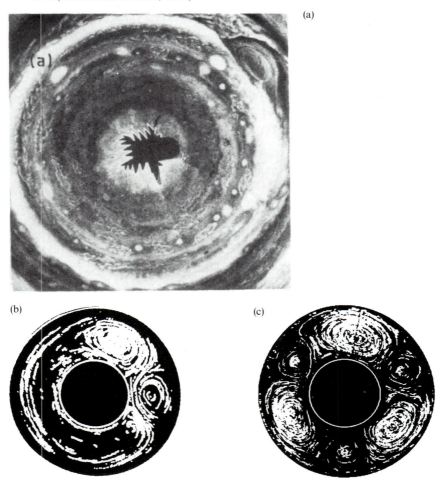

(a)

(b)　　　　　　　　　　　　(c)

13.6 and problem 13.9). 'Blocking' patterns in the earth's atmosphere are also situations in which isolated baroclinic eddies show stability over a substantial period – perhaps several weeks (fig. 13.7). Studies relating these various phenomena are relevant to an understanding of predictability.

13.6   The challenge of climate research

There is enormous scientific and technical challenge in a programme of research directed towards understanding the climate. Combining as it does the need for the highest performance instrumentation, for data management and organization on a very large scale, for the development of complex models

Fig. 13.7. Height in dm of the 50 kPa surface for the Northern Hemisphere on 7 January 1985 illustrating a strong blocking situation over the Atlantic Ocean.

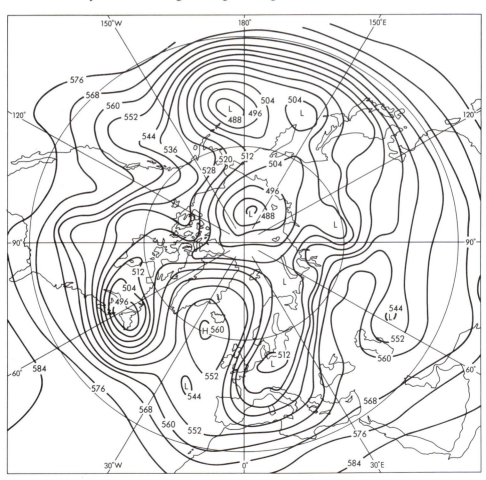

which demand the largest computing capacity available, together with the glamour of a close association with space research, it is virtually unsurpassed as a field for human endeavour. It is also an enterprise about which all of mankind is concerned and in which the whole world can be involved. The problem of understanding the causes of climate change is not one to be solved quickly or easily but contributing to its solution is enormously worthwhile.

Problems

13.1 From equation (4.26), estimate the change in temperature which would occur at the stratopause if the carbon dioxide content of the atmosphere were doubled. In your calculation assume that the ozone concentration does not change.

13.2 Measurements may be made from satellites of the distribution of the flux F_{TA} of net radiation at the top of the atmosphere. Show that the distribution of the flux F_{BA} of net radiation at the bottom of the atmosphere can be written as

$$F_{BA} = F_{TA} - \text{div } T_A - S_A \qquad (13.1)$$

where T_A is the horizontal transport of energy by the atmosphere and S_A is the rate of storage of energy in the atmosphere.

If T_0 and S_0 are respectively the horizontal transport of energy and the rate of storage of energy in the oceans, show that, over the oceans,

$$\text{div } T_0 = F_{BA} - S_O \qquad (13.2)$$

Oort & Vonder Haar (1976) used these equations together with observations of F_{TA}, S_A and S_0 to make an estimate of the transport of heat in the oceans averaged across circles of latitude and its variation with time of year. This is an important quantity in understanding the role of the oceans in climate.

13.3 The intensity of solar radiation incident on the earth and its distribution over the earth varies due to secular changes in the earth's orbit about the sun. These are (fig. 13.8) changes in the eccentricity (e) of the orbit with a period of about 97 000 years, changes in the obliquity (ε) of its axis with a period of about 40 000 years, and changes due to the precession of the longitude (ω) of the perihelion with a period of about 21 000 years. In 1930 Milankovitch (see for instance Mason (1976)) suggested that climatic variations might be linked with these orbital changes, a view which is supported by the fact that strong signals are found in the spectrum of climatic variation at the periods mentioned above associated with orbital variations.

(a) Show from a consideration of the geometry of the earth's orbit that the fractional variation in the incidence of solar radiation at the solstices from its mean value, due to variations in distance from the sun is approximately $2e \sin \omega$. From fig. 13.8 what are the percentage differences in the intensity of solar radiation from its current value at the summer and winter solstices 12 000 years B.P. and 120 000 years B.P.

(b) The zenith angle z of the sun at any given time of day is given by

$$\cos z = \sin \phi \sin \delta + \cos \phi \cos \delta \cos h \qquad (13.3)$$

where ϕ is the latitude, δ the solar declination and h the hour angle.

Integrate the solar radiation incident over a day on a horizontal surface at the top of the atmosphere to find the percentage change in daily insolation at latitude $50°$ at the summer and winter solstices from its current value at 12 000 B.P. and at 120 000 B.P. due to changes in the obliquity factor (see fig. 13.8).

(c) Remembering that for the orbit of a planet around the sun, equal areas are swept out in equal times, show that the difference between the length of the summer half-year T_S and of

Fig. 13.8. Variations of elements of the earth's orbit during the last 250 000 years (from Berger (1982)). The eccentricity e ($=(1-b^2/a^2)^{1/2}$ where a and b are the semi-major axis and semi-minor axis respectively) is shown dashed (left-hand scale), the obliquity ε is the full line (far right-hand scale) and the deviations of the precessional term $\Delta(e \sin \omega)$ from its 1950 A.D. value are shown by the dash–dot line (right-hand scale). ω is the longitude of perihelion relative to the moving vernal equinox.

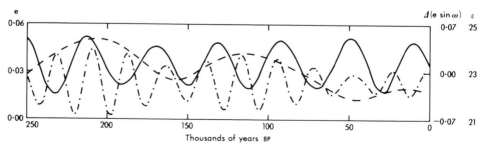

the winter half-year T_W is given by

$$T_S - T_W = \frac{4}{\pi} \, Te \sin \omega \tag{13.4}$$

where T is the length of the year.

13.4 Equation (1.1) for the radiative balance of the earth as a whole can be written in the form

$$\tfrac{1}{4}F(1 - A) = \varepsilon \sigma T_s^4 \tag{13.5}$$

where F is the solar constant, A the albedo, T_s an average surface temperature (~ 280 K) and ε a quantity allowing for the effect of clouds and gaseous components contributing to the greenhouse effect (§ 2.5).

From equation (13.5) show that for constant A and ε, i.e. for no cloud–radiation feedback, a 1% change in F leads to a 0.7 K change in T_s. Show also that a 0.01 change in A leads to a 1 K change in T_s. Assuming that the mixed layer of the oceans with a mean depth of ~ 100 m is involved in these changes in addition to the atmosphere, what is the time constant for the change in T_s resulting in the changes in F or in A.

13.5 Between 18 000 B.P., the time of the last glacial maximum, and 6000 B.P. it is estimated that 40×10^6 km^3 of ice melted from the ice sheets covering the continents. What percentage increase in solar radiation would be required polewards of latitude $45°$ to provide the heat required to melt the ice during the period, assuming all the increase were available for this purpose.

13.6 For a simplified discussion of the cloud–radiation feedback problem, consider a black surface at temperature T_0 illuminated by average solar radiation F such that $F = \sigma T_0^4$. Introduce an idealized cloud in front of the surface such that it is perfectly absorbing and emitting in the infrared but absorbs no solar radiation. The cloud's albedo for solar radiation is A. Consider the radiation balance of the cloud and the new radiation balance of the surface and solve for the cloud temperature T_c and the new surface temperature T_1 in terms of A and T_0. Show that if $A = 0.5$, $T_1 = T_0$, and that the behaviour of $T_1 - T_0$ for different cloud types agrees qualitatively with fig. 13.5. The reason for poor quantitative agreement is that in the simplified calculation the presence of the atmosphere has been ignored.

13.7 The chloro–fluoro–methanes (CFM's) possess absorption bands near 11 μm in the infrared atmospheric window. It has been argued that increase in these compounds in the atmosphere could add to the greenhouse effect and hence alter the mean surface temperature. In the absence of feedback effects, estimate the change in net radiation at the top of the atmosphere which would occur if the combined concentrations of CFM's in the troposphere were 1 part in 10^9, given that the strength of the bands involved near 11μm is ~ 2000 cm^{-1} (atm-cm)$^{-1}$. Hence estimate the change in surface temperature which might occur (cf. (2.15)).

13.8　Volcanic eruptions often injection substantial amounts of dust into the stratosphere. If the particles are of sufficiently small size these layers can persist for many months or even years. Suppose such a layer absorbs 1% of incident solar radiation, while, because of the small size of the particles, having a negligible effect on the long wave-radiation emitted from below. What change in surface temperature might be expected resulting from this dust layer (cf. chapter 2)?

13.9　The character of the flow in a rotating annulus containing fluid in which thermal convection is induced depends on several dimensionless quantities. If the axial and transverse dimensions of the apparatus are H and L respectively, the rotation rate Ω, the fluid density ρ_0, the impressed density contrast $\Delta\rho$, and the kinematic viscosity v, show that the following dimensionless parameters may be formed

$$\Phi = \frac{gH\Delta p}{\Omega^2 \rho_0 L^2} \quad \text{and} \quad \zeta = \frac{\Omega^2 L^5}{v^2 H}$$

where g is the acceleration due to gravity. When ζ is very much greater than about 10^5 in the character of the flow depends mainly on Φ (Read & Hide, 1984 and figs. 10.1 and 13.6).

13.10　Lorenz (1982) has illustrated the problem of atmospheric predictability by investigating a simple non-linear system described by the single first order quadratic difference equation:

$$Y_{n+1} = aY_n - Y_n^2$$

where a is a constant. If $0 \leqslant a \leqslant 4$ and $0 \leqslant Y_0 \leqslant a$, a sequence is generated in which $0 \leqslant Y_n \leqslant a$ for all n.

For $a = 3.75$ and $Y_0 = 1.5$ calculate a table of Y_n for values of n up to 30. Suppose there is an error in Y_0 (representing the initial data) so that $Y_0 = 1.501$. Work out a new sequence of Y_n.

Now set Y_0 back to 1.5 and suppose that a is in error

(representing an error in the model) so that $a = 3.751$. Again work out a new sequence of Y_n.

Finally, to simulate the effect of errors in mathematical procedure, starting from $a = 3.75$ and $Y = 1.5$ compute a series of Y_n in which each value of Y_n is rounded off to four significant figures.

Comment on the difference between the sequences of Y_n. In particular, for each case, how many stages are required for the error to grow by a factor of 10, 100, 1000.

13.11 Various attempts have been made to investigate the stability of the climate system by setting up simple global energy balance models which incorporate parameterizations based on empirical information.

Making use of the empirical relations proposed by Budyko (1969) and Sellers (1969) (see also Wiin Nielsen (1981)) set up a simple global averaged model as follows. Take the incoming radiant energy F_{in} as given by

$$F_{in} = \tfrac{1}{4} F_s (1 - A(T))$$

where F_s is the solar radiation on a surface normal to the solar beam outside the atmosphere and A is the average planetary albedo. A is considered to be influenced mainly by the snow and ice cover and to be a function only of the average surface temperature T, following the empirical relations:

$$A(T) \quad \begin{cases} = A_{max} = 0.85 & T < T_1 = 216\ \text{K} \\[2mm] = A_{max} - \dfrac{A_{max} - A_{min}}{T_2 - T_1}(T - T_1) & T_1 < T < T_2 \\[2mm] = 0.25 & T \geqslant T_2 = 283\ \text{K} \end{cases}$$

Take the outgoing infrared radiation to space as

$$F_{out} = \sigma T^4 [1 - \tfrac{1}{2} \tanh 19 \times 10^{-16} T^6]$$

where the term in brackets represents in a crude way the effect of water vapour, carbon dioxide, and clouds on terrestrial radiation.

Plot as a function of surface temperature T the function $F_{in} - F_{out}$. Show that there are three equilibrium states corresponding to a completely ice covered earth, to partial ice cover and to an ice-free earth. Show also that two of the equilibrium positions are stable with respect to small changes while the other is unstable.

Now reduce the solar constant F_s by 12% and show that only one equilibrium situation remains.

Appendices

1. Some useful physical constants and data on dry air

Avogadro's number	$6.022 \times 10^{26} \, \text{kmol}^{-1}$
Loschmidt number	$2.687 \times 10^{25} \, \text{m}^{-3}$
Boltzmann constant k	$1.381 \times 10^{-23} \, \text{J K}^{-1}$
Planck constant h	$6.6262 \times 10^{-34} \, \text{J s}$
Gas constant R	$8.3143 \, \text{k J K}^{-1} \, \text{k mol}^{-1}$
Stefan–Boltzmann constant σ	$5.670 \times 10^{-8} \, \text{J m}^{-2} \, \text{K}^{-4} \, \text{s}^{-1}$
Velocity of light c	$2.998 \times 10^{8} \, \text{m s}^{-1}$
Ice point	$273.15 \, \text{K}$
Earth's mean radius	$6371 \, \text{km}$
Mean solar angular diameter	31.99 minutes of arc
Standard surface gravity	$9.80665 \, \text{m s}^{-2}$
Standard pressure p_0	$1.01325 \times 10^{5} \, \text{Pa} \, (\equiv 1013.25 \, \text{mb})$

Data on dry air

Apparent molecular weight	28.964
Gas constant for dry air	$287.05 \, \text{J kg}^{-1} \, \text{K}^{-1}$
Specific heats of dry air:	
at constant pressure c_p	$1005 \, \text{J kg}^{-1} \, \text{K}^{-1}$
at constant volume c_v	$718 \, \text{J kg}^{-1} \, \text{K}^{-1}$
Ratio of specific heats γ	1.40
Density of dry air at 273 K and 101.3 kPa pressure	$1.293 \, \text{kg m}^{-3}$
Viscosity (at STP)	$1.73 \times 10^{-5} \, \text{kg m}^{-1} \, \text{s}^{-1}$
Kinematic viscosity (at STP)	$1.34 \times 10^{-5} \, \text{m}^2 \, \text{s}^{-1}$
Thermal conductivity (at STP)	$2.40 \times 10^{-2} \, \text{W m}^{-1} \, \text{K}^{-1}$

Refractive index n of dry air at 101.3 kPa, 273 K, and wavelength of 1 μm
$$= 1 + 289.2 \times 10^{-6}$$

At other wavelengths λ μm Eldén's formula may be used

$$\{n(\lambda) - 1\} \times 10^6 = 64.328 + 29\,498.1\,(146 - \lambda^{-2})^{-1} + 255.4(41 - \lambda^{-2})^{-1}$$

To obtain n at other temperatures and pressures note that $n - 1$ is proportional to density. For air containing water vapour there is a correction which is usually negligible at visible and infrared wavelengths but which becomes significant in the mm wavelength region. (For more information see Penndorf (1957).)

2.   Properties of water vapour

Molecular weight	18.015
Latent heat of fusion at 273 K	$3.34 \times 10^5\,\text{J kg}^{-1}$
Latent heat of vaporization at 273 K	$2.500 \times 10^6\,\text{J kg}^{-1}$
Specific heat of liquid water at 273 K	$4.218 \times 10^3\,\text{J kg}^{-1}\,\text{K}^{-1}$
Specific heat of ice at 273 K	$2.106 \times 10^3\,\text{J kg}^{-1}\,\text{K}^{-1}$
Density of ice at 273 K	$917\,\text{kg m}^{-3}$

Table A2.

After *Smithsonian Meteorological Tables*, Smithsonian Institute, Washington D.C. 1958
Saturation vapour pressure (Pa) *over pure liquid water*

°C	−30	−20	−10	0	+10	+20	+30	+40
0	50.88	125.40	286.27	610.78	1227.2	2337.3	4243.0	7 377.7
+1	55.89	136.64	309.71	656.62	1311.9	2486.1	4492.7	7 780.2
+2	61.34	148.77	334.84	705.47	1401.7	2643.0	4755.1	8 201.5
+3	67.27	161.86	361.77	757.53	1496.9	2808.6	5030.7	8 642.3
+4	73.71	175.97	390.61	812.94	1597.7	2983.1	5320.0	9 103.4
+5	80.70	191.18	421.48	871.92	1704.4	3167.1	5623.6	9 585.5
+6	88.27	207.55	454.51	934.65	1817.3	3360.8	5942.2	10 089
+7	96.49	225.15	489.81	1001.3	1936.7	3564.9	6276.2	10 616
+8	105.38	244.09	527.53	1072.2	2063.0	3779.6	6626.4	11 166
+9	115.00	264.43	567.80	1147.4	2196.4	4005.5	6993.4	11 740

Saturation vapour pressure (Pa) *over pure ice*

°C	-100	-90	-80	-70	-60
0	1.403×10^{-3}	9.672×10^{-3}	5.472×10^{-2}	2.615×10^{-1}	1.080
+1	1.719×10^{-3}	1.160×10^{-2}	6.444×10^{-2}	3.032×10^{-1}	1.236
+2	2.101×10^{-3}	1.388×10^{-2}	7.577×10^{-2}	3.511×10^{-1}	1.413
+3	2.561×10^{-3}	1.658×10^{-2}	8.894×10^{-2}	4.060×10^{-1}	1.612
+4	3.117×10^{-3}	1.977×10^{-2}	1.042×10^{-1}	4.688×10^{-1}	1.838
+5	3.784×10^{-3}	2.353×10^{-2}	1.220×10^{-1}	5.406×10^{-1}	2.092
+6	4.584×10^{-3}	2.796×10^{-2}	1.425×10^{-1}	6.225×10^{-1}	2.380
+7	5.542×10^{-3}	3.316×10^{-2}	1.662×10^{-1}	7.159×10^{-1}	2.703
+8	6.685×10^{-3}	3.925×10^{-2}	1.936×10^{-1}	8.223×10^{-1}	3.067
+9	8.049×10^{-3}	4.638×10^{-2}	2.252×10^{-1}	9.432×10^{-1}	3.476

°C	-50	-40	-30	-20	-10	0
0	3.935	12.83	37.98	103.2	259.7	610.7
+1	4.449	14.36	42.13	113.5	283.7	
+2	5.026	16.06	46.69	124.8	309.7	
+3	5.671	17.94	51.70	137.1	337.9	
+4	6.393	20.02	57.20	150.6	368.5	
+5	7.198	22.33	63.23	165.2	401.5	
+6	8.097	24.88	69.85	181.1	437.2	
+7	9.098	27.69	77.09	198.4	475.7	
+8	10.21	30.79	85.02	217.2	517.3	
+9	11.45	34.21	93.70	237.6	562.3	

3. Atmospheric composition

Table A3. *Normal composition of clean dry air near sea level*

Gas	Volume mixing ratio
Nitrogen (N_2)	0.780 83
Oxygen (O_2)	0.209 47
Argon (Ar)	0.009 34
Carbon dioxide (CO_2)	0.000 33
Neon (Ne)	18.2×10^{-6}
Helium (He)	5.2×10^{-6}
Krypton (Kr)	1.1×10^{-6}
Xenon (Xe)	0.1×10^{-6}
Hydrogen (H_2)	0.5×10^{-6}
Methane (CH_4)	2×10^{-6}
Nitrous oxide (N_2O)	0.3×10^{-6}
Carbon monoxide (CO)	0.1×10^{-6}

For ozone (O_3) distribution see § 5.5.

Water vapour (H₂O)

The above table is for dry air. Average water vapour distribution is shown in fig. A3.1. The water vapour mass mixing ratio of the stratosphere is much more uniform than in the troposphere and is in the range 2 to 5×10^{-6}.

Fig. A3.1. Contours of average water vapour mass mixing ratio in $g\,kg^{-1}$. (After Newell *et al.*, 1972)

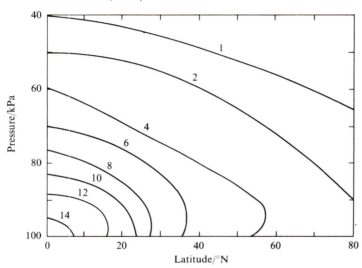

4.   Relation of geopotential to geometric height

Latitude	Geopotential metres (gpm)										
	10 000	20 000	30 000	40 000	50 000	100 000	200 000	300 000	400 000	500 000	600 000
	m	m	m	m	m	m	m	m	m	m	m
0°	10 036	20 104	30 204	40 336	50 500	101 811	206 948	315 577	427 874	544 029	664 243
30°	10 023	20 077	30 163	40 282	50 432	101 672	206 656	315 115	427 225	543 174	663 161
45°	10 009	20 050	30 123	40 228	50 365	101 534	206 363	314 653	426 576	542 318	662 080
60°	9 996	20 024	30 083	40 174	50 297	101 395	206 071	314 191	425 927	541 465	661 000
90°	9 983	19 997	30 043	40 120	50 229	101 256	205 779	313 730	425 280	540 613	659 923

(From *American Institute of Physics Handbook*, McGraw-Hill 1972.)

5.   Model atmospheres (0–105 km)

In the following tables geopotential height, pressure and temperature are plotted for different latitudes and months for typical northern hemisphere conditions. The three quantities are related by the hydrostatic equation (1.4).

Sources of data are Oort & Rasmusson (1971), Labitzke *et al.* (1972), and COSPAR (1972).

For an indication of the difference between the hemispheres see fig. 12.9

The tabulated data of appendix 5 are plotted in fig. 5.1(*a*) and (*b*).

March

Pressure (kPa)	Geopotential height (km) 10°N	Temp. (K)	Geopotential height (km) 40°N	Temp. (K)	Geopotential height (km) 70°N	Temp. (K)
100.0	0.097	299.1	0.127	280.5	0.108	253.2
95.0	0.545	296.6	0.550	278.2	0.503	253.9
90.0	1.016	294.1	0.992	275.8	0.909	254.6
85.0	1.510	291.5	1.454	273.6	1.333	254.8
70.0	3.152	283.1	2.996	266.4	2.770	250.0
50.0	5.864	267.4	5.546	250.5	5.171	236.1
40.0	7.579	256.5	7.146	238.8	6.680	226.7
30.0	9.679	241.4	9.102	225.2	8.549	218.9
20.0	12.41	219.3	11.70	216.4	11.12	218.3
10.0	16.56	196.2	16.10	215.4	15.60	219.5
7.0	18.65	200.9	18.35	215.7	17.96	219.9
5.0	20.62	205.5	20.47	216.1	20.18	220.4
3.0	23.75	217.0	23.70	218.9	23.21	215.6
1.0	30.90	235.0	30.62	226.0	30.27	226.0
0.70	33.38	243.0	33.00	232.0	32.66	229.0
0.50	35.79	250.0	35.27	237.0	34.91	231.0
0.30	39.60	261.0	38.92	248.0	38.39	235.0
0.10	48.29	274.0	47.27	266.0	46.20	249.0
0.70×10^{-1}	51.13	271.0	50.06	267.0	48.81	253.0
0.50×10^{-1}	53.79	267.0	52.66	264.0	51.31	254.0
0.30×10^{-1}	57.67	252.0	56.53	255.0	55.12	252.0
0.10×10^{-1}	65.23	222.0	64.42	234.0	62.90	232.0
0.70×10^{-2}	67.52	214.0	66.81	228.0	65.30	226.0
0.50×10^{-2}	69.64	208.0	69.07	223.0	67.50	222.0
0.30×10^{-2}	72.66	203.0	72.32	217.0	70.78	216.0
0.10×10^{-2}	79.13	196.0	79.13	203.0	77.56	208.0
0.70×10^{-3}	81.15	196.0	81.19	200.0	79.72	206.0
0.50×10^{-3}	83.05	195.0	83.13	197.0	81.75	203.0
0.30×10^{-3}	85.98	194.0	86.06	194.0	84.81	200.0
0.10×10^{-3}	92.24	198.0	92.26	195.0	91.24	202.0
0.70×10^{-4}	94.34	202.0	94.31	198.0	93.37	204.0
0.50×10^{-4}	96.45	208.0	96.36	202.0	95.46	211.0
0.30×10^{-4}	99.76	224.0	99.53	211.0	98.67	223.0
0.10×10^{-4}	108.3	294.0	107.2	256.0	107.2	291.0

June

Pressure (kPa)	Geopotential height (km) 10°N	Temp. (K)	Geopotential height (km) 40°N	Temp. (K)	Geopotential height (km) 70°N	Temp. (K)
100.0	0.098	299.5	0.137	290.4	0.096	277.5
95.0	0.544	296.7	0.571	289.1	0.515	276.8
90.0	1.017	294.3	1.028	287.6	0.953	275.3
85.0	1.513	291.8	1.506	286.0	1.412	273.3
70.0	3.160	283.1	3.112	278.0	2.943	265.6
50.0	5.873	267.0	5.771	261.2	5.488	250.3
40.0	7.587	256.5	7.437	249.5	7.085	238.9
30.0	9.689	241.4	9.482	234.7	9.048	227.7
20.0	12.42	219.1	12.17	219.3	11.73	226.9
10.0	16.60	198.3	16.54	213.7	16.38	229.0
7.0	18,73	203.9	18.79	215.9	18.75	229.5
5.0	20.73	209.2	20.92	218.0	20.99	230.1
3.0	23.91	218.8	24.16	222.9	24.41	231.2
1.0	31.09	238.0	31.26	237.0	31.71	243.0
0.70	33.62	246.0	33.76	244.0	34.28	248.0
0.50	36.06	252.0	36.20	251.0	36.76	255.0
0.30	39.88	260.0	40.04	263.0	40.62	266.0
0.10	48.50	272.0	48.81	276.0	49.53	283.0
0.70×10^{-1}	51.34	271.0	51.66	273.0	52.48	280.0
0.50×10^{-1}	54.00	268.0	54.33	269.0	55.24	276.0
0.30×10^{-1}	57.95	259.0	58.28	256.0	59.28	265.0
0.10×10^{-1}	65.77	227.0	65.97	225.0	67.28	234.0
0.70×10^{-2}	68.09	216.0	68.31	216.0	69.70	224.0
0.50×10^{-2}	70.19	208.0	70.39	208.0	71.81	213.0
0.30×10^{-2}	73.24	201.0	73.41	198.0	74.92	198.0
0.10×10^{-2}	79.61	196.0	79.50	180.0	80.67	166.0
0.70×10^{-3}	81.65	195.0	81.32	176.0	82.35	158.0
0.50×10^{-3}	83.57	195.0	83.04	172.0	83.94	150.0
0.30×10^{-3}	86.45	192.0	85.57	167.0	86.11	147.0
0.10×10^{-3}	92.54	187.0	90.93	171.0	90.90	156.0
0.70×10^{-4}	94.54	188.0	92.67	178.0	92.54	165.0
0.50×10^{-4}	96.44	190.0	94.49	188.0	94.15	176.0
0.30×10^{-4}	99.41	198.0	97.60	207.0	97.15	209.0
0.10×10^{-4}	106.6	243.0	105.3	263.0	105.8	327.0

September

Pressure (kPa)	Geopotential height (km) 10°N	Temp. (K)	Geopotential height (km) 40°N	Temp. (K)	Geopotential height (km) 70°N	Temp. (K)
100.0	0.092	299.4	0.141	291.9	0.076	274.2
95.0	0.538	296.6	0.578	289.9	0.491	272.5
90.0	1.010	294.1	1.037	287.8	0.924	270.8
85.0	1.505	291.5	1.518	285.7	1.378	269.4
70.0	3.149	282.7	3.125	278.1	2.888	262.3
50.0	5.859	266.9	5.789	261.9	5.402	247.4
40.0	7.570	256.3	7.457	250.3	6.981	236.3
30.0	9.670	241.0	9.514	235.8	8.923	224.6
20.0	12.40	218.7	12.21	219.7	11.56	222.8
10.0	16.58	199.7	16.54	211.5	16.11	224.5
7.0	18.72	205.5	18.78	214.3	18.44	224.0
5.0	20.73	211.0	20.89	217.0	20.63	223.6
3.0	23.93	217.6	24.13	221.6	23.95	222.2
1.0	31.07	235.0	31.35	236.0	31.41	232.0
0.70	33.54	241.0	33.83	240.0	33.84	237.0
0.50	35.95	247.0	36.22	246.0	36.20	242.0
0.30	39.70	258.0	39.93	256.0	39.88	252.0
0.10	48.30	272.0	48.48	269.0	48.37	270.0
0.70×10^{-1}	51.14	272.0	51.27	266.0	51.18	271.0
0.50×10^{-1}	53.83	269.0	53.90	260.0	53.84	268.0
0.30×10^{-1}	57.77	259.0	57.67	248.0	57.79	260.0
0.10×10^{-1}	65.67	233.0	65.23	221.0	65.75	235.0
0.70×10^{-2}	68.07	225.0	67.48	214.0	68.17	225.0
0.50×10^{-2}	70.25	218.0	69.57	208.0	70.36	216.0
0.30×10^{-2}	73.44	209.0	72.63	201.0	73.49	204.0
0.10×10^{-2}	80.01	200.0	78.90	190.0	79.71	183.0
0.70×10^{-3}	82.08	199.0	80.87	188.0	81.57	179.0
0.50×10^{-3}	84.03	199.0	82.73	186.0	83.32	175.0
0.30×10^{-3}	87.02	199.0	85.53	187.0	85.93	173.0
0.10×10^{-3}	93.49	207.0	91.71	198.0	91.54	176.0
0.70×10^{-4}	95.81	216.0	93.79	203.0	93.39	178.0
0.50×10^{-4}	98.00	224.0	95.87	207.0	95.19	180.0
0.30×10^{-4}	101.7	247.0	99.06	213.0	97.96	183.0
0.10×10^{-4}	110.9	338.0	106.7	250.0	104.1	188.0

December

Pressure (kPa)	Geopotential height (km) 10°N	Temp. (K)	Geopotential height (km) 40°N	Temp. (K)	Geopotential height (km) 70°N	Temp. (K)
100.0	0.097	299.4	0.133	280.2	0.096	254.5
95.0	0.546	296.7	0.557	278.0	0.491	254.6
90.0	1.017	293.9	1.000	275.7	0.897	255.3
85.0	1.509	291.2	1.465	273.5	1.322	255.6
70.0	3.150	283.0	3.011	266.6	2.764	250.8
50.0	5.863	267.6	5.567	250.9	5.191	236.6
40.0	7.579	256.7	7.170	239.3	6.704	226.9
30.0	9.679	241.2	9.123	225.8	8.556	217.7
20.0	12.41	219.0	11.72	215.8	11.10	214.9
10.0	16.57	196.1	16.09	213.6	15.49	214.3
7.0	18.65	201.7	18.33	214.1	17.83	214.3
5.0	20.62	207.1	20.44	214.7	20.04	214.4
3.0	23.76	216.2	23.64	216.8	22.89	209.0
1.0	30.97	237.0	30.56	225.0	30.36	212.0
0.70	33.47	244.0	32.95	230.0	32.61	215.0
0.50	35.90	251.0	35.22	235.0	34.72	218.0
0.30	39.70	259.0	38.83	245.0	38.04	224.0
0.10	48.27	271.0	47.09	264.0	45.51	245.0
0.70×10^{-1}	51.09	271.0	49.87	266.0	48.14	253.0
0.50×10^{-1}	53.76	269.0	52.46	263.0	50.64	258.0
0.30×10^{-1}	57.71	259.0	56.34	256.0	54.52	256.0
0.10×10^{-1}	65.49	224.0	64.28	235.0	62.47	238.0
0.70×10^{-2}	67.77	214.0	66.68	229.0	64.97	232.0
0.50×10^{-2}	69.87	207.0	68.93	223.0	67.20	228.0
0.30×10^{-2}	72.90	202.0	72.18	216.0	70.55	222.0
0.10×10^{-2}	79.36	201.0	78.96	204.0	77.57	215.0
0.70×10^{-3}	81.46	200.0	81.06	203.0	79.82	214.0
0.50×10^{-3}	83.44	200.0	83.04	202.0	81.92	214.0
0.30×10^{-3}	86.37	193.0	86.07	202.0	85.12	215.0
0.10×10^{-3}	92.43	185.0	92.64	206.0	92.11	218.0
0.70×10^{-4}	94.40	186.0	94.83	208.0	94.42	219.0
0.50×10^{-4}	96.30	189.0	96.95	210.0	96.60	218.0
0.30×10^{-4}	99.27	199.0	100.2	216.0	99.93	216.0
0.10×10^{-4}	106.7	257.0	107.9	248.0	107.1	215.0

March

Geo-potential height (km)	Pressure (kPa) 10°N	Temp. (K)	Pressure (kPa) 40°N	Temp. (K)	Pressure (kPa) 70°N	Temp. (K)
0	101.10	299.6	101.80	281.2	101.40	253.0
1	90.17	294.1	89.91	275.7	88.90	254.6
2	80.22	289.0	79.35	271.0	77.67	252.5
3	71.27	283.8	69.96	266.3	67.78	248.6
4	63.01	278.2	61.31	260.1	58.92	242.8
5	55.66	272.4	53.74	253.9	51.21	237.0
6	49.12	266.5	46.93	247.1	44.23	230.9
7	43.18	260.1	40.82	239.8	38.08	225.3
8	37.76	253.4	35.28	232.8	32.65	221.2
9	32.92	246.2	30.45	225.9	27.94	218.8
10	28.60	238.8	26.08	222.1	23.87	218.5
11	24.65	230.7	22.32	218.7	20.39	218.3
12	21.25	222.6	19.08	216.3	17.46	218.5
13	18.12	216.0	16.30	216.1	14.95	218.8
14	15.33	210.4	13.92	215.8	12.81	219.0
15	12.98	204.8	11.89	215.6	10.97	219.3
16	10.98	199.3	10.15	215.4	9.409	219.5
17	9.280	197.2	8.665	215.5	8.089	219.7
18	7.824	199.5	7.395	215.7	6.954	219.9
19	6.596	201.7	6.311	215.8	5.978	220.1
20	5.561	204.0	5.386	216.0	5.139	220.3
21	4.702	206.8	4.598	216.5	4.355	219.1
22	3.994	210.5	3.926	217.4	3.677	217.5
23	3.393	214.2	3.353	218.3	3.106	215.9
24	2.890	217.5	2.867	219.2	2.650	216.2
25	2.481	219.5	2.459	220.4	2.267	216.9
30	1.140	233.0	1.100	224.0	1.040	226.0
35	0.556	248.0	0.519	237.0	0.494	231.0
40	0.285	262.0	0.259	252.0	0.237	237.0
45	0.150	272.0	0.134	264.0	0.118	247.0
50	0.808×10^{-1}	273.0	0.705×10^{-1}	267.0	0.595×10^{-1}	255.0
55	0.427×10^{-1}	264.0	0.368×10^{-1}	260.0	0.305×10^{-1}	252.0
60	0.218×10^{-1}	242.0	0.188×10^{-1}	245.0	0.153×10^{-1}	240.0
65	0.104×10^{-1}	222.0	0.917×10^{-2}	233.0	0.733×10^{-2}	227.0
70	0.470×10^{-2}	208.0	0.433×10^{-2}	221.0	0.339×10^{-2}	217.0
75	0.202×10^{-2}	200.0	0.196×10^{-2}	211.0	0.153×10^{-2}	210.0
80	0.858×10^{-3}	196.0	0.860×10^{-3}	202.0	0.668×10^{-3}	206.0
85	0.355×10^{-3}	194.0	0.361×10^{-3}	195.0	0.290×10^{-3}	200.0
90	0.149×10^{-3}	195.0	0.150×10^{-3}	192.0	0.123×10^{-3}	200.0
95	0.630×10^{-4}	204.0	0.625×10^{-4}	199.0	0.538×10^{-4}	209.0
100	0.289×10^{-4}	225.0	0.279×10^{-4}	213.0	0.250×10^{-4}	232.0
105	0.146×10^{-4}	263.0	0.134×10^{-4}	239.0	0.129×10^{-4}	270.0

June

Geo-potential height (km)	Pressure (kPa) 10°N	Temp. (K)	Pressure (kPa) 40°N	Temp. (K)	Pressure (kPa) 70°N	Temp. (K)
0	101.10	300.1	101.60	290.8	101.20	277.6
1	90.18	294.3	90.30	287.7	89.47	275.1
2	80.26	289.2	80.07	283.5	78.89	270.3
3	71.33	283.9	70.95	278.5	69.47	265.2
4	63.07	278.1	62.56	272.3	60.87	259.2
5	55.72	272.1	55.12	266.0	53.33	253.2
6	49.18	266.2	48.49	259.6	46.55	246.6
7	43.18	260.1	42.41	252.5	40.48	239.5
8	37.80	253.5	36.95	245.4	34.98	233.6
9	32.97	246.3	32.10	238.1	30.21	227.9
10	28.65	238.8	27.74	231.7	25.98	227.4
11	24.70	230.7	23.85	225.9	22.34	227.1
12	21.29	222.5	20.50	220.2	19.21	227.0
13	18.18	216.2	17.52	218.2	16.55	227.4
14	15.40	211.2	14.96	216.9	14.26	227.9
15	13.04	206.2	12.76	215.6	12.28	228.3
16	11.05	201.3	10.89	214.4	10.58	228.8
17	9.355	199.3	9.298	214.1	9.105	229.1
18	7.910	201.9	7.936	215.1	7.834	229.3
19	6.688	204.6	6.773	216.1	6.740	229.6
20	5.655	207.2	5.780	217.1	5.800	229.8
21	4.790	210.0	4.934	218.1	4.990	230.1
22	4.080	213.0	4.215	219.6	4.299	230.4
23	3.474	216.0	3.601	221.1	3.703	230.7
24	2.961	218.9	3.077	222.6	3.190	231.0
25	2.547	220.9	2.647	224.3	2.754	231.9
30	1.170	235.0	1.200	233.0	1.270	239.0
35	0.577	250.0	0.587	248.0	0.634	249.0
40	0.295	261.0	0.301	263.0	0.324	264.0
45	0.155	269.0	0.160	275.0	0.174	278.0
50	0.829×10^{-1}	273.0	0.862×10^{-1}	275.0	0.944×10^{-1}	283.0
55	0.440×10^{-1}	267.0	0.459×10^{-1}	267.0	0.515×10^{-1}	277.0
60	0.229×10^{-1}	253.0	0.238×10^{-1}	250.0	0.273×10^{-1}	262.0
65	0.113×10^{-1}	230.0	0.116×10^{-1}	229.0	0.140×10^{-1}	243.0
70	0.516×10^{-2}	209.0	0.534×10^{-2}	209.0	0.667×10^{-2}	222.0
75	0.222×10^{-2}	198.0	0.227×10^{-2}	193.0	0.296×10^{-2}	197.0
80	0.933×10^{-3}	196.0	0.907×10^{-3}	179.0	0.115×10^{-2}	169.0
85	0.388×10^{-3}	193.0	0.337×10^{-3}	168.0	0.391×10^{-3}	148.0
90	0.159×10^{-3}	188.0	0.121×10^{-3}	167.0	0.122×10^{-3}	151.0
95	0.645×10^{-4}	188.0	0.460×10^{-4}	191.0	0.433×10^{-4}	186.0
100	0.272×10^{-4}	201.0	0.207×10^{-4}	224.0	0.195×10^{-4}	244.0
105	0.126×10^{-4}	229.0	0.104×10^{-4}	260.0	0.108×10^{-4}	316.0

September

Geo-potential height (km)	Pressure (kPa) 10°N	Temp. (K)	Pressure (kPa) 40°N	Temp. (K)	Pressure (kPa) 70°N	Temp. (K)
0	101.10	299.9	101.70	292.5	100.90	274.5
1	90.10	294.1	90.39	287.9	89.14	270.5
2	80.17	288.8	80.19	283.4	78.47	266.4
3	71.24	283.5	71.07	278.7	68.96	261.6
4	62.98	277.7	62.68	272.7	60.32	255.7
5	55.63	271.9	55.24	266.7	52.76	249.7
6	49.09	266.0	48.61	260.4	45.95	243.2
7	43.09	259.8	42.52	253.4	39.89	236.1
8	37.71	253.1	37.07	246.4	34.40	230.1
9	32.88	245.8	32.24	239.4	29.65	224.5
10	28.57	238.3	27.88	232.9	25.42	223.8
11	24.62	230.1	23.99	226.9	21.79	223.1
12	21.23	221.9	20.64	220.9	18.70	222.9
13	18.11	215.9	17.62	218.2	16.06	223.3
14	15.34	211.4	15.02	216.3	13.79	223.7
15	13.00	206.9	12.80	214.4	11.85	224.0
16	11.02	202.3	10.91	212.5	10.17	224.4
17	9.328	200.8	9.298	212.0	8.727	224.3
18	7.893	203.5	7.927	213.3	7.485	224.1
19	6.678	206.2	6.758	214.6	6.420	223.9
20	5.651	209.0	5.761	215.8	5.507	223.7
21	4.790	211.5	4.913	217.1	4.723	223.4
22	4.083	213.6	4.196	218.5	4.049	223.0
23	3.480	215.6	3.585	220.0	3.472	222.6
24	2.968	217.7	3.062	221.4	2.977	222.2
25	2.551	219.7	2.630	222.7	2.556	222.8
30	1.170	232.0	1.220	233.0	1.230	229.0
35	0.569	245.0	0.592	243.0	0.591	239.0
40	0.289	259.0	0.297	257.0	0.295	252.0
45	0.151	269.0	0.156	267.0	0.153	266.0
50	0.808×10^{-1}	273.0	0.824×10^{-1}	269.0	0.813×10^{-1}	272.0
55	0.431×10^{-1}	267.0	0.433×10^{-1}	257.0	0.431×10^{-1}	266.0
60	0.223×10^{-1}	252.0	0.217×10^{-1}	239.0	0.224×10^{-1}	255.0
65	0.110×10^{-1}	235.0	0.104×10^{-1}	222.0	0.112×10^{-1}	238.0
70	0.520×10^{-2}	219.0	0.466×10^{-2}	207.0	0.530×10^{-2}	218.0
75	0.232×10^{-2}	206.0	0.200×10^{-2}	197.0	0.232×10^{-2}	199.0
80	0.100×10^{-2}	200.0	0.820×10^{-3}	189.0	0.946×10^{-3}	182.0
85	0.424×10^{-3}	199.0	0.331×10^{-3}	186.0	0.360×10^{-3}	173.0
90	0.181×10^{-3}	201.0	0.134×10^{-3}	194.0	0.135×10^{-3}	174.0
95	0.793×10^{-4}	213.0	0.575×10^{-4}	205.0	0.518×10^{-4}	180.0
100	0.378×10^{-4}	236.0	0.260×10^{-4}	216.0	0.208×10^{-4}	185.0
105	0.198×10^{-4}	277.0	0.126×10^{-4}	238.0	0.862×10^{-5}	190.0

December

Geo-potential height (km)	Pressure (kPa) 10°N	Temp. (K)	Pressure (kPa) 40°N	Temp. (K)	Pressure (kPa) 70°N	Temp. (K)
0	101.10	299.9	101.60	280.9	101.30	254.4
1	90.18	294.0	90.00	275.7	88.76	255.3
2	80.20	288.7	79.48	271.1	77.58	253.3
3	71.25	283.7	70.10	266.6	67.75	249.4
4	63.00	278.1	61.45	260.6	58.98	243.5
5	55.65	272.5	53.87	254.3	51.34	237.7
6	49.12	266.7	47.08	247.7	44.38	231.4
7	43.13	260.3	40.96	240.5	38.20	225.4
8	37.76	253.6	35.40	233.5	32.71	220.4
9	32.92	246.2	30.55	226.6	27.95	217.2
10	28.60	238.5	26.17	222.4	23.83	216.1
11	24.65	230.4	22.39	218.5	20.32	215.0
12	21.25	222.3	19.14	215.6	17.35	214.7
13	18.12	215.7	16.33	215.1	14.82	214.6
14	15.34	210.2	13.93	214.6	12.66	214.5
15	12.99	204.7	11.89	214.1	10.81	214.3
16	11.00	199.2	10.14	213.6	9.257	214.3
17	9.292	197.2	8.648	213.8	7.949	214.3
18	7.829	199.9	7.375	214.0	6.826	214.3
19	6.597	202.7	6.289	214.3	5.861	214.3
20	5.558	205.4	5.363	214.5	5.033	214.4
21	4.699	208.2	4.572	215.0	4.212	212.5
22	3.994	211.1	3.897	215.7	3.521	210.7
23	3.395	214.0	3.322	216.3	2.949	209.0
24	2.892	216.7	2.837	217.1	2.505	209.1
25	2.483	218.7	2.436	218.2	2.128	209.3
30	1.150	235.0	1.090	224.0	1.060	211.0
35	0.564	249.0	0.516	235.0	0.479	218.0
40	0.289	260.0	0.255	249.0	0.222	227.0
45	0.151	267.0	0.131	263.0	0.107	244.0
50	0.804×10^{-1}	272.0	0.688×10^{-1}	266.0	0.544×10^{-1}	258.0
55	0.427×10^{-1}	266.0	0.359×10^{-1}	260.0	0.281×10^{-1}	255.0
60	0.221×10^{-1}	252.0	0.183×10^{-1}	247.0	0.142×10^{-1}	244.0
65	0.108×10^{-1}	226.0	0.899×10^{-2}	233.0	0.697×10^{-2}	232.0
70	0.489×10^{-2}	207.0	0.424×10^{-2}	220.0	0.326×10^{-2}	223.0
75	0.210×10^{-2}	200.0	0.191×10^{-2}	211.0	0.150×10^{-2}	217.0
80	0.898×10^{-3}	201.0	0.838×10^{-3}	203.0	0.680×10^{-3}	214.0
85	0.381×10^{-3}	197.0	0.359×10^{-3}	201.0	0.306×10^{-3}	215.0
90	0.157×10^{-3}	187.0	0.155×10^{-3}	204.0	0.138×10^{-3}	217.0
95	0.630×10^{-4}	187.0	0.682×10^{-4}	208.0	0.640×10^{-4}	219.0
100	0.266×10^{-4}	203.0	0.311×10^{-4}	215.0	0.297×10^{-4}	215.0
105	0.215×10^{-4}	240.0	0.149×10^{-4}	232.0	0.138×10^{-4}	212.0

6.   Mean reference atmosphere (110–500 km)

Geometric height (km)	Temp. (K)	Log pressure (kPa)	Mean mol. wt.
110	244	−5.121	26.56
120	335	−5.574	25.45
130	445	−5.892	24.48
140	549	−6.130	23.64
150	635	−6.322	22.91
160	703	−6.487	22.25
170	756	−6.633	21.65
180	798	−6.767	21.10
190	832	−6.890	20.59
200	859	−7.006	20.13
250	940	−7.512	18.33
300	973	−7.947	17.21
350	987	−8.345	16.52
400	993	−8.721	16.02
450	996	−9.070	15.54
500	997	−9.417	14.94

From COSPAR (1972): see chapter 5 for more information.

7.   The Planck function

$$B_{\tilde{v}} = \frac{c_1 \tilde{v}^3}{\exp(c_2 \tilde{v}/T) - 1}$$

where $B_{\tilde{v}}$ is radiance (in $\text{W m}^{-2} \text{ sr}^{-1} (\text{cm}^{-1})^{-1}$) of black body at T K and $\tilde{v}$ wavenumbers (cm^{-1}), c_1 and c_2 are known as first and second radiation constants and have the values

$$c_1 = 1.1911 \times 10^{-8} \text{ W m}^{-2} \text{ sr}^{-1} (\text{cm}^{-1})^{-4}$$

$$c_2 = 1.439 \text{ K} (\text{cm}^{-1})^{-1}$$

$$B_{\lambda} = \frac{c_1}{\lambda^5 (\exp(c_2/\lambda T) - 1)}$$

where B_{λ} is radiance in $\text{W m}^{-2} \text{ sr}^{-1} \text{cm}^{-1}$ of black body at T K and wavelength λ cm, c_1 and c_2 have the same values as above.

Another useful quantity is

$$D = \frac{\int_{\tilde{v}}^{\infty} B_{\tilde{v}} \, d\tilde{v}}{\int_0^{\infty} B_{\tilde{v}} \, d\tilde{v}}$$

A series evaluation of D suitable for numerical computation is

$$D = \frac{15}{\pi^4} \sum_{m=1}^{\infty} m^{-4} \exp(-mv)[\{(mv + 3)mv + 6\}mv + 6]$$

For small values of v, $v < 2\pi$, the following series converges more rapidly

$$D = 1 - \frac{15}{\pi^4} v^3 \left(\frac{1}{3} - \frac{v}{8} + \frac{v^2}{60} - \frac{v^4}{5040} + \frac{v^6}{272\,160} - \frac{v^8}{13\,305\,600} \right)$$

where

$$v = \frac{c_2 \tilde{v}}{T}$$

The wavenumber $\tilde{v}_m$ cm^{-1} of maximum $B_{\tilde{v}}$ is given by Wien's displacement law which is

$$\tilde{v}_m = 1.9609\,T$$

The wavelength λ_m cm of maximum B_λ is given by

$$\lambda_m = 0.289\,79/T.$$

Reference Pivovonsky & Nagel (1961).

8.   Solar radiation

Solar constant (i.e. mean value of total solar radiation incident on surface just outside earth's atmosphere normal to solar beam) $= 1370$ W m^{-2}.

Table A8. *Solar Spectral Irradiance* (after Thekaekara, 1973)

$\lambda =$ wavelength, μm.

$I_\lambda =$ solar spectral irradiance averaged over small bandwidth centred at λ (W m^{-2} nm^{-1}).

$D_{0-\lambda} =$ percentage of the solar constant associated with wavelengths shorter than λ.

λ	I_λ	$D_{0-\lambda}$	λ	I_λ	$D_{0-\lambda}$	λ	I_λ	$D_{0-\lambda}$
0.115	7×10^{-6}	1×10^{-4}	0.43	1.660	12.47	0.90	0.902	63.37
0.14	3×10^{-5}	5×10^{-4}	0.44	1.833	13.73	1.00	0.757	69.49
0.16	2.3×10^{-4}	6×10^{-4}	0.45	2.031	15.14	1.2	0.491	78.40
0.18	0.00127	1.6×10^{-3}	0.46	2.092	16.65	1.4	0.341	84.33
0.20	0.0108	8.1×10^{-3}	0.47	2.059	18.17	1.6	0.248	88.61
0.22	0.0582	0.05	0.48	2.100	19.68	1.8	0.161	91.59
0.23	0.0675	0.10	0.49	1.975	21.15	2.0	0.104	93.49
0.24	0.0638	0.14	0.50	1.966	22.60	2.2	0.080	94.83
0.25	0.0718	0.19	0.51	1.906	24.01	2.4	0.063	95.86
0.26	0.132	0.27	0.52	1.856	25.38	2.6	0.049	96.67
0.27	0.235	0.41	0.53	1.865	26.74	2.8	0.039	97.31
0.28	0.225	0.56	0.54	1.805	28.08	3.0	0.031	97.83
0.29	0.488	0.81	0.55	1.747	29.38	3.2	0.0229	98.22
0.30	0.520	1.21	0.56	1.716	30.65	3.4	0.0168	98.50
0.31	0.698	1.66	0.57	1.734	31.91	3.6	0.0137	98.72
0.32	0.840	2.22	0.58	1.737	33.18	3.8	0.0112	98.91
0.33	1.072	2.93	0.59	1.721	34.44	4.0	0.0096	99.06
0.34	1.087	3.72	0.60	1.687	35.68	4.5	0.0060	99.34
0.35	1.107	4.52	0.62	1.622	38.10	5.0	0.0038	99.51
0.36	1.081	5.32	0.64	1.563	40.42	6.0	0.0018	99.72
0.37	1.190	6.15	0.66	1.505	42.66	7.0	0.0010	99.82
0.38	1.134	7.00	0.68	1.445	44.81	8.0	6.0×10^{-4}	99.88
0.39	1.112	7.82	0.70	1.386	46.88	10.0	2.5×10^{-4}	99.94
0.40	1.447	8.73	0.72	1.331	48.86	15.0	4.9×10^{-5}	99.98
0.41	1.773	9.92	0.75	1.251	51.69	20.0	1.5×10^{-5}	99.99
0.42	1.770	11.22	0.80	1.123	56.02	50.0	4×10^{-7}	100.00

Fig. A8.1. Solar radiation curves. Shaded areas show absorption by vertical path of whole atmosphere by constituents shown. (From Air Force Cambridge Research Laboratories, 1965)

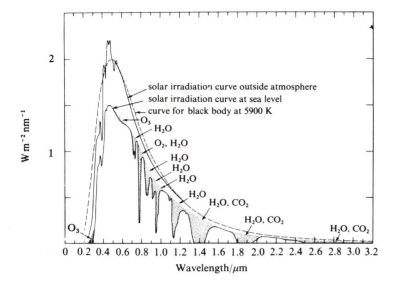

9. Absorption of solar radiation by oxygen and ozone

Table A9.1. *Absorption cross-sections per molecule of* O_2 *and* O_3 *in* cm^2

λ (nm)	O_2	O_3
150	16.8×10^{-18}	4.7×10^{-19}
160	5.32×10^{-18}	1.1×10^{-18}
170	1.86×10^{-18}	8.2×10^{-19}
180	15.0×10^{-24}	7.4×10^{-19}
190	15.4×10^{-24}	5.1×10^{-19}
200	13.0×10^{-24}	2.9×10^{-19}
210	9.6×10^{-24}	4.5×10^{-19}
220	6.4×10^{-24}	1.9×10^{-18}
230	3.1×10^{-24}	4.6×10^{-18}
240	1.0×10^{-24}	8.0×10^{-18}
250	3.0×10^{-25}	10.8×10^{-18}
260		10.7×10^{-18}
270		7.8×10^{-18}
280		3.7×10^{-18}
290		1.3×10^{-18}
300		4×10^{-19}
310		1×10^{-19}
320		3×10^{-20}
330		7×10^{-21}
340		2×10^{-21}
450		0.2×10^{-21}
500		1.0×10^{-21}
550		3.3×10^{-21}
600		4.7×10^{-21}
650		2.7×10^{-21}
700		0.9×10^{-21}
750		0.3×10^{-21}

Values for O_2 are from Ditchburn & Young (1962), for O_3 below 300 nm from Inn & Tanaka (1953), above 300 nm from Vigroux (1953). The cross-sections for O_2 include the effect of Rayleigh scattering.

Table A9.2. *Solar absorption due to ozone by all ultraviolet and visible bands for various path lengths along solar beam*

Ozone path length (cm STP)	Rate of absorption of solar radiation (mW cm^{-2})
0.0001	0.0195
0.0005	0.0928
0.0010	0.1772
0.0050	0.6530
0.0100	0.9766
0.0500	1.848
0.1000	2.345
0.2000	3.005
0.3000	3.523
0.4000	3.979
0.5000	4.400
0.6000	4.796
0.7000	5.175
0.8000	5.541
0.9000	5.895
1.0000	6.241
2.0000	9.360
3.0000	12.096
4.0000	14.564
5.0000	16.818
6.0000	18.887
7.0000	20.795
8.0000	22.559
9.0000	24.195
10.0000	25.714

From Kennedy (1964).

Absorption by near infrared O$_2$ bands (after Houghton, 1963)

For atmospheric paths where collision broadening is dominant, absorption by the O$_2$ bands at 0.76 μm and 1.27 μm follows the square-root law (4.39) with $\sum (s_i \gamma_{oi})^{1/2} = 2.3$ cm^{-1} (g cm^{-2})$^{-1/2}$ for 0.76 μm band and $= 0.07$ cm^{-1} (g cm^{-2})$^{-1/2}$ for 1.27 μm band.

10.   Spectral band information

The following tables list values of $\sum s_i$ (columns S) and $\sum (s_i \gamma_{0i})^{1/2}$ (columns R) at different temperatures over different wavenumber intervals of various infrared bands of water vapour, carbon dioxide and ozone, where s_i is the strength of the ith line in $cm^{-1} (g\ cm^{-2})^{-1}$ and γ_{0i} is the collision broadened half-width in cm^{-1} of the ith line at standard pressure (101.3 kPa). The data are from McClatchey *et al.* (1973). More explanation is given in chapter 4.

In the tables, each number is given by a number between 1 and 10 expressed to four significant figures followed by an exponent of ten, e.g. the value for S at 220 K for water vapour between 0 and $25\ cm^{-1}$ is 3.615×10^3.

Water vapour

Temperature		220 K				260 K				300 K			
Wavenumber interval		S		R		S		R		S		R	
0	25	3.615	3	3.191	1	2.487	3	2.677	1	1.801	3	2.309	1
25	50	3.706	4	2.097	2	2.971	4	1.829	2	2.426	4	1.619	2
50	75	1.187	5	4.251	2	9.374	4	3.797	2	7.622	4	3.433	2
75	100	2.176	5	5.404	2	1.768	5	4.944	2	1.469	5	4.563	2
100	125	1.973	5	4.853	2	1.854	5	4.772	2	1.711	5	4.642	2
125	150	2.290	5	4.858	2	2.063	5	4.698	2	1.863	5	4.529	3
150	175	2.955	5	4.867	2	2.754	5	4.847	2	2.552	5	4.783	2
175	200	7.944	4	2.498	2	9.435	4	2.804	2	1.078	5	3.04	2
200	225	2.450	5	4.231	2	2.460	5	4.259	2	2.395	5	4.230	2
225	250	1.259	5	2.649	2	1.407	5	2.844	2	1.494	5	2.981	2
250	275	7.900	4	1.600	2	9.516	4	1.760	2	1.047	5	1.867	2
275	300	6.289	4	2.085	2	8.495	4	2.368	2	1.023	5	2.563	2
300	325	4.341	4	1.581	2	6.570	4	1.883	2	8.630	4	2.119	2
325	350	1.870	4	1.085	2	3.193	4	1.346	2	4.645	4	1.562	2
350	375	9.650	3	8.448	1	1.822	4	1.126	2	2.934	4	1.383	2
375	400	4.413	3	6.273	1	9.153	3	8.692	1	1.638	4	1.110	1
400	425	3.004	3	3.724	1	5.209	3	5.274	1	8.319	3	6.876	1
425	450	7.356	2	2.322	1	1.337	3	3.232	1	2.700	3	4.365	1
450	475	1.475	3	3.374	1	2.598	3	4.652	1	4.084	3	5.994	1
475	500	3.264	2	9.622	0	4.893	2	1.413	1	8.023	2	2.023	1
500	525	5.422	2	1.654	1	1.079	3	2.435	1	1.856	3	3.340	1
525	550	3.074	2	1.088	1	5.415	2	1.453	1	8.219	2	1.870	1

Water vapour(continued)

Wavenumber interval		220 K S		220 K R		260 K S		260 K R		300 K S		300 K R	
550	575	1.358	2	7.994	0	2.318	2	1.083	1	3.601	2	1.429	1
575	600	1.635	2	8.508	0	3.686	2	1.272	1	6.631	2	1.713	1
600	625	4.782	1	3.981	0	1.026	2	5.926	0	1.955	2	8.264	0
625	650	5.678	1	4.855	0	1.379	2	7.300	0	2.681	2	9.932	0
650	675	2.575	1	3.175	0	4.857	1	4.496	0	8.300	1	5.975	0
675	700	1.779	1	2.660	0	5.154	1	4.345	0	1.124	2	6.186	0
700	725	9.761	0	1.876	0	2.430	1	3.040	0	5.154	1	4.427	0
725	750	5.737	0	1.524	−1	1.543	1	2.453	0	3.479	1	3.555	0
750	775	1.764	0	7.205	0	4.776	0	1.196	0	1.145	1	1.817	0
775	800	5.477	0	1.567	0	1.410	1	2.536	0	2.939	1	3.621	0
1200	1250	1.682	1	5.706	1	3.394	1	8.364	0	6.166	1	1.166	1
1250	1300	8.281	1	1.562	1	1.760	2	2.274	1	3.411	2	3.133	1
1300	1350	6.019	2	4.324	1	1.271	3	5.869	1	2.253	3	7.493	1
1350	1400	5.161	3	9.781	1	8.007	3	1.206	2	1.098	4	1.408	2
1400	1450	1.043	4	1.313	2	1.357	4	1.499	2	1.629	4	1.653	2
1450	1500	3.198	4	2.609	2	3.526	4	2.763	2	3.739	4	2.888	2
1500	1550	6.466	4	4.780	2	6.573	4	5.012	2	6.640	4	5.210	2
1550	1600	4.774	4	2.673	2	4.392	4	2.559	2	4.062	4	2.462	2
1600	1650	4.252	4	2.640	2	3.667	4	2.450	2	3.209	4	2.306	2
1650	1700	8.557	4	4.408	2	7.739	4	4.220	2	7.057	4	4.059	2
1700	1750	4.263	4	3.173	2	4.412	4	3.252	2	4.463	4	3.298	2
1750	1800	2.301	4	1.906	2	2.553	4	2.062	2	2.781	4	2.192	2

Start	End	V1	E1	V2	E2	V3	E3	V4	E4	V5	E5	V6	E6
1800	1850	8.472	3	1.021	2	9.912	3	1.150	2	1.134	4	1.270	2
1850	1900	2.955	3	5.064	1	3.575	3	5.746	1	4.097	3	6.408	1
1900	1950	1.896	3	4.285	1	2.581	3	5.030	1	3.186	3	5.690	1
1950	2000	4.045	2	2.033	1	6.707	2	2.546	1	9.549	2	3.028	1
2000	2050	1.109	2	1.084	1	2.157	2	1.511	1	3.504	2	1.948	1
2050	2100	3.618	1	6.195	0	7.955	1	9.316	0	1.425	2	1.274	1
2800	2900	5.145	0	9.590	0	6.051	0	1.023	1	7.034	0	1.084	1
2900	3000	2.932	1	1.207	1	4.680	1	1.543	1	6.665	1	1.874	1
3000	3100	4.137	2	4.243	1	4.351	2	4.425	1	4.472	2	4.574	1
3100	3200	5.370	2	4.656	1	4.888	2	4.682	1	4.537	2	4.785	1
3200	3300	7.944	2	6.302	1	7.602	2	6.460	1	7.371	2	6.683	1
3300	3400	3.752	2	4.683	1	4.641	2	5.672	1	5.745	2	6.672	1
3400	3500	7.461	2	7.408	1	1.105	3	9.470	1	1.559	3	1.164	2
3500	3600	8.166	3	2.762	2	1.198	4	3.267	2	1.623	4	3.710	2
3600	3700	6.589	4	5.927	2	6.819	4	5.947	2	6.873	4	5.926	2
3700	3800	8.352	4	7.355	2	7.880	4	7.280	2	7.525	4	7.222	2
3800	3900	1.084	5	7.192	2	1.025	5	7.026	2	9.669	4	6.854	2
3900	4000	7.588	3	1.718	2	1.112	4	2.109	2	1.483	4	2.461	2
4000	4100	3.194	2	2.943	1	3.707	2	3.378	1	4.262	2	3.877	1
4100	4200	4.760	1	1.307	1	6.812	1	1.505	1	8.804	1	1.671	1
4900	5000	3.722	0	5.244	0	6.931	0	6.874	0	1.120	1	8.453	0
5000	5100	4.905	1	1.678	1	7.132	1	2.052	1	1.029	2	2.455	1
5100	5200	8.698	2	7.783	1	1.270	3	9.159	1	1.692	3	1.040	2
5200	5300	7.303	3	1.779	2	7.231	3	1.755	2	7.053	3	1.722	2
5300	5400	1.170	4	2.295	2	1.069	4	2.254	2	9.958	3	2.228	2
5400	5500	1.008	4	2.080	2	1.031	4	2.082	2	1.031	4	2.065	2

Water vapour(continued)

Wavenumber interval		220 K S		R		260 K S		R		300 K S		R	
5500	5600	5.709	2	5.023	1	9.270	2	6.271	1	1.343	3	7.381	1
5600	5700	5.333	1	1.261	1	6.648	1	1.552	1	8.475	1	1.885	1
5700	5800	1.016	1	4.586	0	1.410	1	5.517	0	1.790	1	6.341	0
6600	6700	2.291	1	1.123	1	2.970	1	1.308	1	3.802	1	1.485	1
6700	6800	3.071	2	4.013	1	3.454	2	4.328	1	3.748	2	4.573	1
6800	6900	5.157	2	5.203	1	4.634	2	5.106	1	4.246	2	5.082	1
6900	7000	7.808	2	7.108	1	7.838	2	7.610	1	8.108	2	8.122	1
7000	7100	1.014	3	9.688	1	1.453	3	1.127	2	1.887	3	1.251	2
7100	7200	5.269	3	1.660	2	5.367	3	1.684	2	5.362	3	1.695	2
7200	7300	8.422	3	2.167	2	7.630	3	2.078	2	7.016	3	2.000	2
7300	7400	8.410	3	2.191	2	8.462	3	2.209	2	8.403	3	2.202	2
7400	7500	2.821	2	4.518	1	4.046	2	5.205	1	5.630	2	5.883	1
7500	7600	8.358	1	2.354	1	8.864	1	2.484	1	9.295	1	2.585	1
7600	7700	2.571	1	9.890	1	3.108	1	1.113	1	3.581	1	1.209	1
8200	8300	4.875	0	5.427	0	5.328	0	5.704	0	5.681	0	5.903	0
8300	8400	1.001	1	7.840	1	9.111	0	7.571	0	8.405	0	7.370	0
8400	8500	1.586	1	9.678	1	1.487	1	9.590	0	1.410	1	9.523	0
8500	8600	7.987	0	8.424	0	1.183	1	1.021	1	1.722	1	1.201	1
8600	8700	1.326	2	2.629	1	1.700	2	2.989	1	2.020	2	3.266	1
8700	8800	3.629	2	3.702	1	3.358	2	3.591	1	3.104	2	3.500	1
8800	8900	8.407	2	6.932	1	7.732	2	6.740	1	7.189	2	6.584	1
8900	9000	3.489	2	4.387	1	3.951	2	4.639	1	4.300	2	4.795	1

9000	9100	2.886	1	1.530	1	3.189	1	1.622	1	3.703	1	1.717	1
9100	9200	9.974	0	7.007	0	1.050	1	7.366	1	1.095	1	7.587	0
9200	9300	3.973	0	3.162	0	4.662	0	3.462	0	5.182	0	3.663	0
10 200	10 300	1.679	1	8.723	0	1.685	1	8.976	1	1.686	1	9.221	0
10 300	10 400	3.038	1	1.411	1	3.123	1	1.487	1	3.381	1	1.559	1
10 400	10 500	8.384	1	2.540	1	9.901	1	2.694	1	1.105	2	2.782	1
10 500	10 600	2.084	2	3.420	1	2.061	2	3.435	2	2.028	2	3.424	1
10 600	10 700	3.267	2	4.055	1	2.889	2	3.812	2	2.578	2	3.602	1
10 700	10 800	1.252	2	2.719	1	1.466	2	2.993	2	1.646	2	3.204	1
10 800	10 900	1.165	1	9.701	0	1.286	1	1.004	1	1.389	1	1.021	1
10 900	11 000	2.751	1	1.556	1	2.736	1	1.559	1	2.707	1	1.549	1
11 000	11 100	3.378	1	1.276	1	3.031	1	1.213	1	2.753	1	1.156	1
11 100	11 200	2.448	1	1.008	1	2.606	1	1.057	1	2.717	1	1.087	1

Carbon dioxide

425	450	6.891	−5	6.845	−3	7.322	−4	2.206	−2	4.060	−3	5.116	−2
450	475	1.681	−3	7.073	−2	1.293	−2	1.921	−1	5.681	−2	3.945	−1
475	500	9.310	−4	7.316	−2	7.474	−3	2.069	−1	3.416	−2	4.395	−1
500	525	1.952	−2	2.870	−1	1.029	−1	6.952	−1	3.502	−1	1.337	0
525	550	2.785	−1	1.215	0	1.086	0	2.514	0	2.945	0	4.375	0
550	575	5.495	−1	2.404	0	2.684	0	5.438	0	8.844	0	1.009	1
575	600	5.331	1	1.958	1	1.324	2	3.365	1	2.609	2	5.118	1
600	625	5.196	2	5.804	1	9.797	2	8.290	1	1.562	3	1.099	2
625	650	7.778	3	2.084	2	9.657	3	2.530	2	1.149	4	3.015	2
650	675	8.746	4	7.594	2	8.597	4	8.171	2	8.526	4	8.908	2
675	700	2.600	4	2.635	2	2.693	4	3.049	2	2.776	4	3.547	2

Carbon dioxide (continued)

Wavenumber interval	220 K				260 K				300 K			
	S		R		S		R		S		R	
700–725	1.232	3	8.387	1	2.339	3	1.237	2	3.756	3	1.684	2
725–750	2.042	2	2.852	1	4.099	2	4.572	1	6.906	2	6.670	1
750–775	7.278	0	6.239	0	2.693	1	1.328	1	7.086	1	2.349	1
775–800	1.337	0	2.765	0	4.864	0	5.651	0	1.283	1	9.858	0
800–825	3.974	−1	8.897	−1	1.482	0	1.805	0	3.866	0	3.065	0
825–850	1.280	−2	3.198	−1	9.151	−2	8.611	−1	3.878	−1	1.779	0
850–875	2.501	−3	1.506	−1	1.948	−2	4.394	−1	9.093	−2	9.715	−1
875–900	3.937	−3	1.446	−1	2.629	−2	4.097	−1	1.211	−1	9.027	−1
900–925	2.320	−2	3.543	−1	1.539	−1	9.242	−1	6.447	−1	1.900	0
925–950	2.859	−1	8.677	−1	1.249	0	1.912	0	3.691	0	3.425	0
950–975	4.707	−1	1.024	0	1.699	0	1.948	0	4.290	0	3.150	0
975–1000	1.906	−1	6.602	−1	8.960	−1	1.431	0	2.791	0	2.519	0
1000–1025	4.418	−2	2.610	−1	2.257	−1	5.667	−1	7.527	−1	1.024	0
1025–1050	3.458	−1	7.736	−1	1.338	0	1.629	0	3.600	0	2.862	0
1050–1075	1.086	0	2.036	0	3.865	0	3.971	0	9.836	0	6.537	0
1075–1100	2.904	−2	5.006	−1	1.792	−1	1.196	0	7.142	−1	2.307	0
1800–1825	4.192	−5	5.444	−3	2.041	−4	1.155	−2	6.397	−4	1.976	−2
1825–1850	3.066	−3	7.370	−2	1.104	−2	1.329	−1	2.887	−2	2.064	−1
1850–1875	4.567	−2	4.378	−1	9.630	−2	6.529	−1	1.739	−1	9.040	−1
1875–1900	3.064	−1	1.604	0	6.218	−1	2.186	0	1.102	0	2.829	0
1900–1925	5.309	0	4.130	0	5.982	0	4.660	0	6.603	0	5.160	0
1925–1950	3.358	0	3.153	0	3.301	0	3.220	0	3.273	0	3.328	0
1950–1975	3.763	−1	7.438	−1	5.232	−1	9.260	−1	6.793	−1	1.112	0
1975–2000	1.557	−2	2.048	−1	4.208	−2	3.207	−1	8.888	−2	4.495	−1

ν_1	ν_2						
2000	2025	1.585×10^{-1}	5.893×10^{-1}	1.928×10^{-1}	6.977×10^{-1}	2.422×10^{-1}	8.162×10^{-1}
2025	2050	3.456×10^{0}	2.788×10^{0}	5.657×10^{0}	3.290×10^{0}	8.228×10^{0}	3.797×10^{0}
2050	2075	2.981×10^{1}	7.642×10^{0}	3.119×10^{1}	8.149×10^{0}	3.248×10^{1}	8.755×10^{0}
2075	2100	2.889×10^{1}	1.172×10^{1}	3.002×10^{1}	1.317×10^{1}	3.181×10^{1}	1.480×10^{0}
2100	2125	8.862×10^{-1}	2.368×10^{0}	1.709×10^{0}	3.612×10^{0}	2.870×10^{0}	5.074×10^{1}
2125	2150	3.189×10^{-1}	1.244×10^{0}	6.608×10^{-1}	2.025×10^{0}	1.191×10^{0}	3.001×10^{0}
2150	2175	1.886×10^{-2}	4.146×10^{-1}	6.865×10^{-2}	8.317×10^{-1}	1.893×10^{-1}	1.415×10^{0}
2175	2200	1.168×10^{-2}	2.821×10^{-1}	7.551×10^{-2}	7.959×10^{-1}	3.256×10^{-1}	1.772×10^{0}
2200	2225	9.504×10^{-1}	2.866×10^{0}	4.393×10^{0}	6.472×10^{0}	1.417×10^{1}	1.213×10^{1}
2225	2250	2.217×10^{2}	3.000×10^{1}	4.106×10^{2}	4.487×10^{1}	6.610×10^{2}	6.270×10^{1}
2250	2275	4.566×10^{3}	1.134×10^{2}	4.829×10^{3}	1.341×10^{2}	5.097×10^{3}	1.653×10^{2}
2275	2300	7.964×10^{3}	2.011×10^{2}	9.287×10^{3}	2.675×10^{2}	1.283×10^{4}	3.628×10^{2}
2300	2325	1.055×10^{5}	5.880×10^{2}	1.482×10^{5}	7.552×10^{2}	1.930×10^{5}	9.312×10^{2}
2325	2350	5.587×10^{5}	1.206×10^{3}	5.412×10^{5}	1.267×10^{3}	5.281×10^{5}	1.348×10^{3}
2350	2375	6.819×10^{5}	1.182×10^{3}	6.796×10^{5}	1.225×10^{3}	6.793×10^{5}	1.278×10^{3}
2375	2400	1.256×10^{4}	8.873×10^{1}	2.320×10^{4}	1.249×10^{2}	3.640×10^{4}	1.617×10^{2}
2400	2425	7.065×10^{-2}	3.404×10^{-1}	2.524×10^{-1}	6.465×10^{-1}	6.360×10^{-1}	1.040×10^{0}
2425	2450	8.522×10^{-2}	4.236×10^{-1}	2.955×10^{-1}	7.922×10^{-1}	7.297×10^{-1}	1.261×10^{0}
2450	2475	1.346×10^{-2}	2.405×10^{-1}	5.986×10^{-2}	4.569×10^{-1}	1.927×10^{-1}	7.630×10^{-1}
2475	2500	4.765×10^{-2}	4.248×10^{-1}	4.907×10^{-2}	4.598×10^{-1}	5.684×10^{-2}	5.231×10^{-1}
3100	3200	1.324×10^{-1}	9.836×10^{-1}	1.593×10^{-1}	1.167×10^{0}	1.955×10^{-1}	1.349×10^{0}
3200	3300	7.731×10^{-2}	4.900×10^{-1}	1.116×10^{-1}	6.244×10^{-1}	1.531×10^{-1}	7.563×10^{-1}
3300	3400	1.232×10^{0}	2.952×10^{0}	1.440×10^{0}	3.374×10^{0}	1.666×10^{0}	3.792×10^{0}
3400	3500	5.159×10^{0}	7.639×10^{0}	8.430×10^{0}	1.124×10^{1}	1.284×10^{1}	1.601×10^{1}
3500	3600	4.299×10^{3}	1.914×10^{2}	5.007×10^{3}	2.321×10^{2}	5.757×10^{3}	2.779×10^{2}
3600	3700	1.543×10^{4}	3.245×10^{2}	1.577×10^{4}	3.505×10^{2}	1.618×10^{4}	3.831×10^{2}
3700	3800	1.649×10^{4}	2.722×10^{2}	1.634×10^{4}	3.012×10^{2}	1.646×10^{4}	3.364×10^{2}

Carbon dioxide (continued)

Wavenumber interval		220 K S		220 K R		260 K S		260 K R		300 K S		300 K R	
3800	3900	1.180	−1	9.535	−1	3.642	−1	1.510	0	8.943	−1	2.260	0
3900	4000	1.464	−2	2.601	−1	1.409	−2	2.532	−1	1.372	−2	2.488	−1
4000	4100	1.251	−2	2.021	−1	2.380	−2	2.672	−1	4.219	−2	3.384	−1
4500	4600	9.976	−3	2.559	−1	1.357	−2	3.022	−1	1.789	−2	3.432	−1
4600	4700	2.185	−1	1.916	0	2.290	−1	2.026	0	2.433	−1	2.132	0
4700	4800	2.040	0	6.475	0	3.437	0	8.577	0	5.522	0	1.105	1
4800	4900	1.197	2	3.112	1	1.220	2	3.314	1	1.249	2	3.525	1
4900	5000	4.829	2	5.759	1	4.880	2	6.340	1	4.977	2	7.006	1
5000	5100	8.778	1	2.012	1	9.629	1	2.231	1	1.058	2	2.460	1
5100	5200	8.346	1	1.804	1	8.699	1	2.063	1	9.187	1	2.332	1
5200	5300	8.518	−2	8.474	−1	1.227	−1	1.122	0	1.734	−1	1.390	0
5300	5400	4.951	−1	1.597	0	4.943	−1	1.639	0	4.969	−1	1.676	0

Ozone

Wavenumber interval		220 K S		220 K R		260 K S		260 K R		300 K S		300 K R	
0	25	5.791	2	1.352	2	4.678	2	1.187	2	3.861	2	1.054	2
25	50	2.012	3	2.923	2	1.746	3	2.759	2	1.535	3	2.603	2
50	75	2.056	3	2.556	2	2.032	3	2.563	2	1.950	3	2.518	2
75	100	8.957	2	1.572	2	1.125	3	1.802	2	1.292	3	1.955	2
100	125	1.834	2	6.041	1	3.153	2	8.070	1	4.554	2	9.783	1
125	150	1.666	1	1.545	1	4.401	1	2.518	1	8.686	1	3.519	1
900	925	2.622	−4	5.784	−3	2.045	−3	1.549	−2	8.913	−3	3.121	−2
925	950	5.162	−2	1.838	−1	1.839	−1	4.040	−1	4.844	−1	7.212	−1

950	975	2.220	1	3.345	1	6.569	1	5.873	1	1.621	2	9.121	1
975	1000	1.623	3	3.521	2	3.107	3	4.834	2	5.054	3	6.124	2
1000	1025	3.446	4	1.397	3	3.862	4	1.541	3	4.157	4	1.654	3
1025	1050	7.260	4	2.207	3	6.542	4	2.218	3	6.013	4	2.246	3
1050	1075	6.183	4	1.652	3	6.330	4	1.726	3	6.398	4	1.774	3
1075	1100	9.484	2	2.204	2	1.002	3	2.233	2	1.051	3	2.244	2
1100	1125	1.074	3	1.755	2	1.027	3	1.711	2	9.843	2	1.666	2
1125	1150	9.191	2	1.661	2	1.010	3	1.710	2	1.089	3	1.735	2
1150	1175	3.287	2	8.363	1	3.654	2	8.689	1	3.901	2	8.827	1
1175	1200	2.647	1	9.530	0	3.468	1	1.070	1	4.166	1	1.148	1

Bibliography

The following books and review articles are suggested for further reading. They are approximately in the order of the chapters to which they are particularly relevant (and which are indicated by bracketed numbers at the end of the reference) except that those general works which cover most of the field are listed first.

Matveev, L. T. (1967). *Fundamentals of General Meteorology: Physics of the Atmosphere*. Israel Programme for Scientific Translations, Jerusalem.

Wallace, J. M. & Hobbs, P. V. (1977). *Atmospheric Science: An Introductory Survey*. Academic Press.

Hess, S. C. (1959). *Introduction to Theoretical Meteorology*. Holt, New York.

Hunten, D. M. (1972). 'Composition and structure of planetary atmospheres.' *Space Sci. Rev.*, **12**, 539–99 (1).

Hunten, D. M., Colin, L., Donahue, T. M. & Moroz, V. I. (1983). *Venus*. University of Arizona Press (1, 7, 10, 12).

Goody, R. M. (1964). *Atmospheric Radiation*. Oxford University Press (2, 4).

Kondratye'v, K. (1969). *Radiation in the Atmosphere*. Academic Press (2, 4).

Brunt, D. (1939). *Physical and Dynamical Meteorology*. Cambridge University Press (3).

Iribarne, J. V. & Godson, W. L. (1973). *Atmospheric Thermodynamics*. Reidel, Dordrecht (3).

Dutton, J. R. & Johnson, D. R. (1967). 'The theory of available potential energy.' *Adv. in Geophys.*, **12**, 333–436 (3, 10).

Corby, G. (ed.) (1970). *The Global Circulation of the Atmosphere*. R. Met. Soc. London (3, 5, 9, 10, 11).

Coulson, K. L. (1975). *Solar and Terrestrial Radiation*. Academic Press (4).

Paltridge, G. W. & Platt, C. M. R. (1976). *Radiative Processes in Meteorology and Climatology*. Elsevier (4).

Akasofu, S. & Chapman, S. (1972). *Solar Terrestrial Physics*. Oxford University Press (5).

Murgatroyd, R. (1970). 'The physics and dynamics of the stratosphere and mesosphere.' *Rep. Prog. Phys.*, **33**, 817–80 (5).

Murata, H. (1974). 'Wave motions in the atmosphere and related ionospheric phenomena.' *Space Sci. Rev.*, **16**, 461–525 (5, 8).

COSPAR (1972). *COSPAR International Reference Atmosphere (CIRA)*. Akademie-Verlag, Berlin (5).

World Meteorological Organisation (1986). WMO/NASA Ozone Assessment, WMO Geneva (5, 10).

Royal Society, London (1980). *The Middle Atmosphere as Observed by Balloons, Rockets and Satellites* (5, 12).

Mason, B. J. (1974). *The Physics of Clouds.* Oxford University Press (6).

Mason, B. J. (1975). *Clouds, Rain and Rainmaking,* 2nd ed. Cambridge University Press (6).

Rogers, R. R. (1976). *A Short Course in Cloud Physics.* Pergamon (6).

van de Hulst, H. C. (1957). *Light Scattering by Small Particles.* Wiley (6).

Feigel'son, E. M. (1966). *Light and Heat Radiation in Stratus Clouds.* Israel Programme for Scientific Translations, Jerusalem (6).

Hansen, J. E. & Travis, L. D. (1974). 'Light scattering in planetary atmospheres.' *Space Sci. Rev.,* **16**, 527–610 (6).

Pedlosky, J. (1971). 'Geophysical fluid dynamics.' *Lectures in Applied Maths.,* **13**, 1–60 (7, 8, 10).

Holton, J. R. (1979). *An Introduction to Dynamic Meteorology.* Academic Press (7, 8, 9, 10, 11).

Charney, J. (1973). 'Planetary fluid dynamics.' In *Dynamic Meteorology,* ed. P. Morel. Reidel, Dordrecht (7, 8, 10).

Gill, A. E. (1982). *Atmosphere–Ocean Dynamics.* Academic Press (7, 8, 9, 10).

Scorer, R. S. (1978). *Environmental Aerodynamics.* Ellis Horwood (7, 8, 9).

Haltiner, G. J. & Williams, R. T. (1980). *Numerical Prediction and Dynamic Meteorology.* John Wiley (7, 11).

Hoskins, B. J. & Pearce, R. P. (eds.) (1983). *Large Scale Dynamical Processes in the Atmosphere.* Academic Press (7, 8, 10).

Chapman, S. & Lindzen, R. S. (1970). *Atmospheric Tides,* Reidel, Dordrecht, Holland (8).

Eckart, C. (1960). *Hydrodynamics of Oceans and Atmospheres.* Pergamon (8).

Lumley, J. L. & Panofsky, H. A. (1964). *The Spectrum of Atmospheric Turbulence.* Interscience (9).

Pasquill, F. & Smith, F. B. (1983). *Atmospheric Diffusion.* Ellis Horwood, Third edn. (9).

Priestley, C. H. B. (1959). *Turbulent Transfer in the Lower Atmosphere.* University of Chicago Press (9).

Monteith, J. (1973). *Principles of Environmental Physics.* Edward Arnold (9).

Lorenz, E. N. (1967). *The Nature and Theory of the General Circulation of the Atmosphere.* World Meteorological Organization, Geneva (10).

Palmén, E. & Newton, C. W. (1969). *Atmospheric Circulation Systems.* Academic Press (10).

Starr, V. P. (1968). *Physics of Negative Viscosity Phenomena.* McGraw-Hill (10).

Hide, R. & Mason, P. J. (1975). Sloping convection in a rotating fluid.' *Adv. in Phys.,* **24**, 47–100 (10).

Holton, J. R. (1975). *The Dynamic Meteorology of the Stratosphere and Mesosphere.* American Met. Soc. (10).

Houghton, J. T. (ed.) (1984). *The Global Climate.* Cambridge University Press (10, 11, 13).

Richardson, L. F. (1922). *Weather Prediction by Numerical Processes.* Cambridge University Press (reprinted by Dover Publications, 1966) (11).

GARP Publication Series No. 14 (1974). *Modelling for the First GARP Global Experiment* (can be obtained from World Meteorological Organization, Geneva) (11).

Thompson, P. D. (1961). *Numerical Weather Analysis and Prediction*. Macmillan (11).

GARP Publication Series No. 17 (1976). *Numerical Methods used in Atmospheric Models*. World Meteorological Organization, Geneva (11).

Dickinson, A. & Temperton, C. (1984). *The Operational Numerical Weather Prediction Model*. Meteorological Office Met. 0 11 Technical Note No. 183 (11).

Slingo, A. (ed.) (1985). *Handbook of the Meteorological Office 11-layer atmospheric general circulation model*, Vol. 1. Meteorological Office DCTN 29 (11).

Houghton, J. T., Taylor, F. W. & Rodgers, C. D. (1984). *Remote Sensing of Atmospheres*. Cambridge University Press (12).

Browning, K. A. (ed.) (1982). *Nowcasting* Academic Press (12).

Lamb, H. H. (1972). *Climate Present, Past and Future*. Methuen (13).

American Meteorological Society (1968). *Causes of Climatic Change*, ed. J. M. Mitchell, Jr. Meteorological Monographs, **8**, no. 30 (13).

GARP Publication Series No. 16 (1975). *The Physical Basis of Climate and Climate Modelling* (can be obtained from World Meteorological Organization, Geneva) (13).

Berger, A., Imbrie, J., Hays, J., Kukla, G. & Saltzman, B. (eds.) *Milankovitch and Climate. NATO* (1984). *ASI Series*. D. Reidel Dordrecht, 2 vols. (13).

References to works cited in the text

Abel, P. G. *et al.* (1970). *Proc. R. Soc. Lond.*, A **320**, 35–55.

Air Force Cambridge Research Laboratories (1965). *Handbook of Geophysics and Space Environments.*

Atkins, M. J. & Woodage, M. J. (1985). *Meteor. Mag.*, **114**, 227–33.

Andrews, D. G., Holton, J. R., Leovy, C. B. (1987). *Middle Atmosphere Dynamics*, Academic Press.

Batchelor, G. (1967). *An Introduction to Fluid Dynamics.* Cambridge University Press.

Bengtsson, L. *et al.* (1982). *Bull. Amer. Met. Soc.*, **63**, 21–43.

Berger, A. L. (1982). *Quaternary Res.*, **9**, 139–67.

Brewer, A. W. (1949). *Q.J.R. Met. Soc.*, **75**, 351–63.

Briggs, G. & Taylor, F. (1982). *Photographic Atlas of the Planets.* Cambridge University Press.

Browning, K. A. (ed.) (1982). *Nowcasting.* Academic Press.

Budyko, M. I. (1969). *Tellus*, **21**, 611–19.

Chapman, S. (1930). *Mem. R. Met. Soc.*, **3**, 103.

Charney, J. G. (1973). In *Dynamical Meteorology*, ed. P. Morel. Reidel, Dordrecht, Holland.

Charney, J. G. & Drazin, P. G. (1961). *J. Geophys. Res.*, **66**, 83–109.

Charney, J., Fjörtoft, R. & von Neumann, J. (1950). *Tellus*, **2**, 237–54.

Clough, S. A., Grahame, N. S. & O'Neill, A. (1985). *Q.J.R. Met. Soc.*, **111**, 335–58.

COSPAR (1972). *COSPAR International Reference Atmosphere (CIRA).* Akademie-Verlag, Berlin.

Crane, A. J. (1977). Unpublished D. Phil. Thesis, University of Oxford.

Crutzen, P. J. (1971). *J. Geophys. Res.*, **76**, 7311–27.

Cullen, M. J. P., Norbury, J. & Purser, R. J. (1986). *Quart. J.R. Met. Soc.* to be published.

Cullen, M. J. P. & Purser, R. J. (1984). *J. Atmos. Sci.*, **41**, 1477–97.

Defant, A. (1961). *Physical Oceanography*, **1**, 489–92. Pergamon Press.

Dickinson, A. & Temperton, C. (1984). Meteorological Office Met 0 11 TN 183.

Ditchburn, R. W. & Young, P. A. (1962). *J. Atmos. and Terres. Phys.*, **24**, 127–39.

Dobson, G. M. B. (1914). *Q. J. R. Met. Soc.*, **40**, 123–35.

Dütsch, H. U. (1968). *Q. J. R. Met. Soc.*, **94**, 483–97.

Dütsch, H. U. (1971). *Adv. in Geophys.*, **15**, 219–322.

Dutton, J. A. & Johnson, D. R. (1967). *Adv. in Geophys.*, **12**, 333–436.

Eady, E. J. (1949). *Tellus*, **1**, 33–52.

Eisberg, R. M. (1961). *Fundamentals of Modern Physics*. John Wiley.

Eyre, J. R., Brownscombe, J. L. & Allam, R. J. (1984). *Meteor. Mag.*, **113**, 264–71.

Fjörtoft, R. (1953). *Tellus.*, **5**, 225.

Gadd, A. J. (1985). *Meteor. Mag.*, **114**, 222–6.

Fritts, D. C. (1984). *Rev. Geophys. & Space Phys.*, **22**, 275–308.

Gerbier, N. & Berenger, M. (1961). *Q. J. R. Met. Soc.*, **87**, 13–23.

Golding, B. W. (1984). *Meteor. Mag.*, **113**, 288–302.

Gill, A. E. (1982). *Atmosphere–Ocean Dynamics*. Academic Press.

Goldman, A. (1968). *J. Quant. Spectrosc. and Radiat. Transfer*, **8**, 829–31.

Goody, R. M. & Walker, J. C. G. (1972). *Atmospheres*. Prentice-Hall.

Hanel, R. A., Schlachman, B., Rodgers, D. & Vanous, D. (1971). *Appl. Opt.*, **10**, 1376–82.

Harwood, R. S. (1975). *Q. J. R. Met. Soc.*, **101**, 75–94.

Harwood, R. S. & Pyle, J. A. (1975). *Q. J. R. Met. Soc.*, **101**, 723–48.

Hays, J. D., Imbrie, J. & Shackleton, M. J. (1976). *Science*, **194**, 1121–32.

Hide, R. (1974). *Proc. R. Soc. London A*, **336**, 63–84.

Hide, R. & Mason, P. J. (1975). *Adv. in Phys.*, **24**, 47–100.

Hide, R. (1982). *Q. Jl. R. astr. Soc.*, **23**, 220–35.

Holloway, J. L. & Manabe, S. (1971). *Mon. Weather Rev.*, **99**, 335–70.

Hoskins, B. J. (1975). *J. Atmos. Sci.*, **32**, 233–42.

Hoskins, B. J., McIntyre, M. E. & Robertson, A. W. (1986). *Q. J. R. Met. Soc.* **111**, 877–946.

Houghton, J. T. (1963). *Q. J. R. Met. Soc.*, **89**, 319–31.

Houghton, J. T. (1972). *Bull. Amer. Met. Soc.*, **53**, 27–8.

Houghton, J. T. & Smith, S. D. (1966). *Infra-red Physics*. Oxford University Press.

Houghton, J. T. & Smith, S. D. (1970). *Proc. R. Soc. London A*, **320**, 23–33.

Houghton, J. T., Taylor, F. W. & Rodgers, C. D. (1984). *Remote Sounding of Atmospheres*. Cambridge University Press.

Houghton, J. T. (1978). *Quart. J. R. Met. Soc.*, **104**, 1–29.

Inn, E. C. Y. & Tanaka, Y. (1953). *J. Opt. Soc. Amer.*, **43**, 870.

Ingersoll, A. P. (1969). *J. Atmos. Sci.*, **26**, 1191–8.

Jeans, J. H. (1940). *Kinetic Theory of Gases*. Cambridge University Press.

Jeffreys, H. (1925). *Q. J. R. Met. Soc.*, **51**, 347–56.

Julian, P. R., Washington, W. M., Hembree, L. & Ridley, C. (1970). *J. Atmos. Sci.*, **27**, 376–87.

Kennedy, J. S. (1964). *Energy Generation Through Radiation Processes in the Lower Stratosphere*. M.I.T. Report, no. 11, Boston.

Krueger, A. J., Heath, D. E. & Mateer, C. L. (1973). *Pure and Appl. Geophys.*, **106–8**, 1254–63.

Labitzke, K. and collaborators (1972). *Climatology of the Stratosphere – the*

Northern Hemisphere Part I. Institut für Meteorologie der Freien Universität
   Berlin. Meteorologische Abhanglungen Band 100/Heft 4.

Lamb, H. H. (1966). *The Changing Climate: Selected Papers.* Methuen, London.

Leovy, C. B. (1964). *J. Atmos. Sci.,* **21**, 327–41.

Leovy, C. B. (1969). *Appl. Opt.,* **8**, 1279–86.

Lindzen, R. S. (1971). In *Mesospheric Models and Related Experiments,* ed. G.
   Fiocco. Reidel, Dordrecht, Holland.

Lorenz, E. (1955). *Tellus,* **7**, 157–69.

Lorenz, E. (1967). *The Nature and Theory of the General Circulation of the
   Atmosphere.* World Meteorological Organization, Geneva.

Lorenz, E. (1975). *Climate Predictability* in GARP Publication Series No. 16
   World Meteorological Organization, Geneva.

Lorenz, E. (1982). In *Problems and Prospects in Long and Medium Range
   Forecasting,* pp. 1–20. European Centre for Medium Range Weather Forecasts,
   Reading.

McClatchey, R. A., Benedict, W. S., Clough, S. A., Burch, D. E., Calfee, R. F., Fox,
   K., Rothman, L. S. & Garing, J. S. (1973). *AFCRL Atmospheric Absorption Line
   Parameters Compilation.* Air Force Cambridge Research Laboratories
   Environment Research Papers no. 434.

McClatchey, R. A. & Selby, J. E. A. (1972). *Atmospheric Transmittance 7–30 µm.* Air
   Force Cambridge Research Laboratories Environmental Research Papers no.
   419.

Malkmus, W. (1967). *J. Opt. Soc. Amer.,* **57**, 323–9.

Manabe, S. (1969). *Mon. Weather Rev.,* **97**, 739–805.

Manabe, S. & Wetherald, R. T. (1967). *J. Atmos. Sci.,* **24**, 241–59.

Mason, B. J. (1976). *Q. J. R. Met. Soc.,* **102**, 473–98.

Matveev, L. T. (1967). *Fundamentals of General Meteorology: Physics of the
   Atmosphere.* Israel Programme for Scientific Translations, Jerusalem.

Milne, E. (1930). Reprinted in *Selected Papers on the Transfer of Radiation,* ed. D.
   H. Menzel. Dover, 1966.

Mintz, Y. (1961). In *The Atmospheres of Mars & Venus.* Publication 944. Ad Hoc
   Panel on Planetary Atmospheres of Space Science Board, National Academy of
   Sciences, Washington D.C.

Mitchell, J. M. Jnr (1977). 'The changing climate.' In *Energy and Climate, Studies
   in Geophysics.* National Academy of Science, Washington D.C.

Monin, A. S. (1973). In *Dynamical Meteorology,* ed. P. Morel. Reidel, Dordrecht,
   Holland.

Murgatroyd, R. J. (1969). In *The Global Circulation of the Atmosphere* (ed. Corby),
   R. Met. Soc. London, pp. 159–95.

Murgatroyd, R. J. & Singleton, F. (1961). *Q. J. R. Met. Soc.,* **87**, 125–35.

Newell, R. E., Kidson, J. W., Vincent, D. G. & Boer, G. J. (1972). *The General
   Circulation of the Tropical Atmosphere,* vol. 1. M.I.T. Press, Boston.

Newton, C. W. (1970). In *The Global Circulation of the Atmosphere,* ed. G. Corby.
   R. Met. Soc. London.

Oort, A. H. & Peixoto, J. P. (1974). *J. Geophys. Res.,* **18**, 2705–19.

Oort, A. H. & Rasmusson, E. M. (1971). *Atmospheric Circulation Statistics.* Professional Paper no. 5, National Oceanographic & Atmospheric Administration, Washington, D.C.

Oort, A. H. & Vonder Haar, T. H. (1976). *J. Phys. Oceang.*, **6**, 781–800.

Palmén, E. & Newton, C. W. (1969). *Atmospheric Circulation Systems.* Academic Press.

Palmer, T. N., Schutts, G. J. & Swinbank, R. (1986). *Q. J. R. Met. Soc.*, to be published.

Penndorf, R. (1957). *J. Opt. Soc. Amer.*, **47**, 176.

Pivovonsky, M. & Nagel, M. R. (1961). *Table of Black-body Functions.* Macmillan.

Plumb, R. A. (1982). *Aust. Met. Mag.*, **30**, 107–21.

Rasool, S. I. & DeBergh, G. (1970). *Nature, Lond.*, **226**, 1037–9.

Read, P. L. & Hide, R. (1984). *Nature*, **308**, 45–8.

Rodgers, C. D. (1967). *Q. J. R. Met. Soc.*, **93**, 43–54.

Rutherford, D. E. (1959). *Fluid Mechanics.* Oliver & Boyd.

Schofield, J. T. & Taylor, F. W. (1983). *Q. J. R. Met. Soc.*, **109**, 57–80.

Sellers, W. O. (1969). *J. Appl. Met.*, **8**, 392–400.

Slingo, A. (ed.) (1985). DCTN 29 Meteorological Office, Bracknell.

Slingo, J. M. (1980). *Quart. J. R. Met. Soc.*, **106**, 747–70.

Smith, W. L. (1976). *Proceedings of Study Conference on Four-dimensional Assimilation.* World Meteorological Organization, Geneva.

Staelin, D. H., Kunzi, K. G., Pettyjohn, R. L., Poon, R. K. L. & Wilcox, R. W. (1976). *J. Appl. Met.*, **15**, 1204–14.

Starr, V. P. (1968). *Physics of Negative Viscosity Phenomena.* McGraw-Hill.

Stephens, G. L. & Webster, P. J. (1981). *J. Atmos. Sci.*, **38**, 235–47.

Thekaekara, M. P. (1973). *Solar Energy*, **14**, 109–27, Pergamon.

Uccellini, L. W., Keyser, D., Brill, K. F. & Wash, C. H. (1985). *Mon. Weather Rev.*, **113**, 962–88.

Vonder Haar, T. & Suomi, V. (1971). *J. Atmos. Sci.*, **28**, 305–14.

Vigroux, E. (1953). *Annales de Phys.*, **8**, 709.

Webster, W. J. Jr, Wilheit, T. T., Chang, T. C., Gloersen, P. & Schmugge, T. J. (1975). *Sky and Telescope*, **49**, 14–16.

Wiin-Nielsen (1981). *Atmospheric–Ocean*, **19**, 89–215.

Wilheit, T. T., Chang, A. T. C., Rao, M. S. V., Rodgers, F. B. & Theon, J. S. (1977). *J. Appl. Met.*, **16**, 551–60.

Williams, A. P. (1971). Unpublished D.Phil. Thesis, University of Oxford.

World Meteorological Organization (1986). WMO/NASA Ozone Assessment, WMO, Geneva.

Answers to problems

and hints to their solution

1.1 Earth: 276 K,
   Venus: 326 K,
   Mars: 224 K,
   Jupiter: 121 K.

1.2 $10.8\ \mathrm{W\ m^{-2}}$.

1.3 $p(z) = p_0\left(1 - \dfrac{\alpha z}{T_0}\right)^{Mg/\alpha R}$

1.4 (1) 19.5 km; (2) 14.2 km (use the result of q 1.3).

1.5 3.07%: $\sim 20\%$.

1.6 Mass of atmosphere $= 5.27 \times 10^{18}$ kg.
   Thermal capacity of atmosphere $= 5.30 \times 10^{21}\ \mathrm{J\ K^{-1}}$.
   Thermal capacity of the oceans $= 5.69 \times 10^{24}\ \mathrm{J\ K^{-1}}$.

1.7 Using the value for c_p for CO_2 of $834\ \mathrm{J\ K^{-1}\ kg^{-1}}$ ($\approx \frac{9}{2}R$) and for H_2 of $14\,450\ \mathrm{J\ K^{-1}\ kg^{-1}}$ ($\approx \frac{7}{2}R$),
   Mars: 4.5
   Venus: 10.6
   Jupiter: $1.8\ \mathrm{K\ km^{-1}}$.

1.8 $9.64\ \mathrm{K\ km^{-1}}$ for damp air compared with $9.76\ \mathrm{K\ km^{-1}}$ for dry air.

1.9 Solve equation iteratively or graphically: radius of balloon 13.3 m.

1.10 Period of oscillation $= 9.8$ minutes.

2.2 $\chi_0^* = 1.814$; $\phi = 183\ \mathrm{W\ m^{-2}}$; Temperature discontinuity 20.7 K.

2.4 $\tau = 0.14$ days.

2.5 Consider rotation of planet around the sun, as well as about its own axis; 1 Venus solar day $= 116$ earth days, $\tau \approx 0.44$ Venus days.

2.6 $\chi_0^* \approx 224$.

3.1 $T^* = T(1 + m/\varepsilon)/(1 + m)$

The required saturation mixing ratios may conveniently be read from the tephigram chart:

at 273 K, $T^* - T = 0.64$ K,

at 290 K, $T^* - T = 2.14$ K,

and

at 300 K, $T^* - T = 4.11$ K.

3.5 (1) 30 kPa.

(2) Ascent is everywhere stable for dry air. It is stable for saturated air(if the mixing ratio is increased so that the air is saturated) above 78 kPa.

(3) At 100 kPa, 8.3 g kg^{-1}; at 50 kPa, 0.75 g kg^{-1}.

(4) 2 K.

(5) 95 kPa.

(6) 65 kPa.

3.7 Work done on 1st stage of ascent = 7 J

Energy released in 2nd stage   = 65 J.

3.8 The temperature of the final mixture is 18°C, and hence, for condensation to occur, mixing ratio of the mixture = 13.1 g kg^{-1}. Relative humidity = 93.7%.

3.9 (2) The approximation is that the process is replaced by one which is isentropic.

$\bar{\theta} = 28.6$°C

$\bar{m} = 18.5$ g kg^{-1}.

Condensation level $\simeq 94$ kPa.

3.10 Energy released = 2800 J kg^{-1}.

1.8 W m^{-2}:

Average solar input 245 W m^{-2}.

3.11 4×10^6 J m^{-2}.

4.1 At 0.3 μm, 0.637; at 0.6 μm, 0.061; at 1 μm, 0.008.

4.2 7.5%.

4.3 7.10^8 m^{-3}.

4.4 Turbidity coefficients, 0.74, 0.15.

Extinctions, 0.826, 0.304, 0.146.

4.6 $h_m = F_s e^{-1/3} g / 2 c_p p_m$, 0.88 K hr^{-1}.

4.7 If $B_v(T)$ is proportional to solar flux, i.e. ignoring all dynamics, changes of ozone concentration etc., temperature difference is ~ 7 K.

4.8 (1) 0.568 kPa, (2) 2.13 kPa, (3) 0.133 kPa.

4.15 0.97 (0.98 for CO_2, 0.99 for H_2O);

0.65 (0.79 for CO_2, 0.82 for H_2O).

4.17 Mass mixing ratio of $H_2O = 0.622e/p$ (3.11).

(1) 0.924, (2) 0.92.

For last part use (4.20). Multiply pathlength by 1.66 in transmission term. Assume spectral interval $\sim 500\,\mathrm{cm}^{-1}$ wide. Answer $1.0\,\mathrm{K\,day}^{-1}$.

4.19 To perform the integral over s write it as

$$\underset{\varepsilon \to 0}{\mathrm{Lt}}\; N_0 \int \int_\varepsilon^\infty \frac{1}{s} \exp\left(-\frac{s}{\sigma}\right) ds [1 - \exp(-sf\rho l)]\, dv$$

$$= N_0 \int \left[\int_\varepsilon^\infty \frac{1}{s} \exp\left(-\frac{s}{\sigma}\right) ds - \int_{\varepsilon'}^\infty \frac{1}{s'} \exp\left(-\frac{s'}{\sigma}\right) ds' \right] dv$$

where $\varepsilon' = \varepsilon(1 + \sigma f\rho l)$.

5.1 Assume mixing ratio of He at 120 km same as at the surface (Appendix 3) and that oxygen completely dissociated at 120 km. For temperature of 800 K, 770 km; for 1000 K, 940 km; for 1400 K, 1260 km.

5.2 If T is orbital period, $\Delta T/T = -1.3 \times 10^{-7}$ per orbit.

5.3 $8.3 \times 10^{-10}\,\mathrm{m\,s}^{-1}$.

5.5 $1.36 \times 10^{11}\,\mathrm{m}^{-3}$; $2.2 \times 10^{11}\,\mathrm{m}^{-2}\,\mathrm{s}^{-1}$.

5.7 The resulting temperature profile rises sharply above 120 km, then flattens out so that above 400 km the temperature is substantially constant at about 1250 K.

5.8 Total potential energy, $130\,\mathrm{J\,m}^{-2}$.

Power absorbed (from information in problem 5.7), $80\,\mathrm{J\,m}^{-2}$. Thus we might expect $\Delta T/T$, ~ 0.6 compared with ~ 0.3 from fig. 5.4.

5.9 $(J_3 + k_2 n_2 m_2)^{-1}$; $\dfrac{J_3 + k_2 n_2 n_M}{4n_2 (J_2 J_3 k_2 k_3 n_M)^{1/2}}$.

First time constant: 0.76 s at 40 km; 0.05 s at 30 km; and 2×10^{-3} s at 20 km. Second time constant: 1.6×10^5 s at 40 km; 1.3×10^6 s at 30 km; 1.3×10^7 s at 20 km.

5.10 $0.4\,\mathrm{K\,day}^{-1}$.

5.12 0.023.

5.13 $g_2/g_1 = 2$ for $15\,\mu\mathrm{m}$ CO_2 band; cooling rate$\simeq 5\,\mathrm{K\,day}^{-1}$.

5.14 $90\,\mathrm{K\,day}^{-1}$.

5.15 Transmission at line centre$\simeq 0.66$.

5.16 7.5×10^9; 2.2×10^{-5}.

5.17 Ratio about 1:100.

5.18 With tropopause temperature 196 K, volume mixing ratio 9.10^{-6}.

6.3 (1) 26.5 min; (2) 7.34 hr.

6.4 91.9 days; 22 hours; 13.2 min.

6.5 22.9 min; $-9.6°C$.

6.6 $A = 2/3$, $\tau = 1/3$.

6.9 $A = 0.6635$, $\tau = 0.3353$, $(1 - A - \tau) = 5.97 \times 10^{-3}$.

6.10 19.56.

6.11 $\beta = 0.6138$, $\chi_0^* = 34.4$.

6.12 $\beta = 0.0266$, with $\chi_0^* = 4$, $\tau = 0.067$.

6.13 0.9972.

7.1 $\Omega^2 R = 3.457 \times 10^{-3} g$: 5.95 arc minutes.

7.3 Magnitudes of terms (cm s^{-2}):

$$\frac{du}{dt}, \frac{dv}{dt} \simeq 10^{-2}, \frac{uv \tan \phi}{a}, \frac{u^2 \tan \phi}{a} \simeq 10^{-3},$$

$$2\Omega u \sin \phi, \, 2\Omega v \sin \phi \simeq 10^{-1}, \, 2\Omega w \cos \phi \simeq 10^{-4}.$$

uw/a, $vw/a \simeq 10^{-6}$.

7.4 0.39 kPa/100 km.

7.5 7.29 m s^{-1}.

7.6 1 day at 30° latitude.

7.7 0.194. On Mars, 0.20; on Venus, 47.1.

7.8 $\sim 10\%$.

7.11 $g = g_0 a^2 (a + z)^{-2}$.

7.12 19 m s^{-1}.

7.13 Consider pressures on either side of the frontal surface, p_1, p_2, and write

$$\delta p_1 = \frac{\partial p_1}{\partial x} \delta x + \frac{\partial p_1}{\partial z} \delta z$$

and similarly for δp_2. At front, $p_1 = p_2$, $\delta p_1 = \delta p_2$ and $\tan \alpha = \delta z / \delta x$. $\alpha = 0.41°$ at 30° latitude.

7.15 Temperature will fall.

7.16 $V_{20\,kPa} = 17.9$ m s^{-1} at 45° latitude.

7.17 Average temperature gradient, 1 K per 100 km; wind speed, 33.5 m s^{-1}.

7.25 Transformations of the form derived in q. 7.24 are applied to the continuity equation.

7.26 For scalar ϕ and vector $\mathbf{F}$, curl $\phi \mathbf{F} = \phi$ curl $\mathbf{F} + \nabla \phi \wedge \mathbf{F}$. Also curl $\nabla \phi = 0$, and g is the gradient of a scalar potential.

8.4 1.88×10^{-2} s^{-1}.

8.5 $\lambda_h \simeq 10$ km, $\lambda_v \simeq 1$ km, horizontal phase speed $\simeq 16$ m s^{-1}.

8.6 $\omega_a = \omega_B$ if there is 2.2 K km^{-1} temperature inversion.

8.8 The kinetic energy density is comprised of terms of form $\frac{1}{2}\rho u^2$: 0.14 m s^{-1} if $\bar{T} = 260$ K.

8.9 (1) 3.34 km; (2) 4.78 km.

8.10 $(v_g)_h = \omega_B \dfrac{m^2}{(m^2 + k^2)^{3/2}}$, $(v_g)_v \equiv \propto \dfrac{\omega_B km}{(m^2 + k^2)^{3/2}}$.

8.12 2.67 m s^{-1}.

8.13 2.48 m s^{-1}.

8.14 $v_g = \bar{u} + \dfrac{\beta(k^2 - l^2)}{(k^2 + l^2)^2}$.

8.15 0.02; 0.01.

8.16 ζ may be approximated to the vorticity of geostrophic wind to an approximation of order Ro.

8.18 1.5×10^{-5}: 3×10^{-5}.

8.19 The method is as for q 8.10 or 8.14.

8.20 The method is as for q 8.2 or 8.11.

9.3 The effect of buoyancy must be considered. We expect $K_E \simeq K$; K_Q will be different.

9.5 If $\zeta_g = 10^{-5} \text{ s}^{-1}$, $H = 10 \text{ km}$, $K = 10 \text{ m}^2 \text{ s}^{-1}$, $w_d \simeq 7 \text{ mm s}^{-1}$. Spin down time $\simeq 4$ days.

Decay time under damping from eddy viscosity but ignoring vertical motion and secondary circulation $\simeq H^2/K$, ~ 100 days.

10.1 (a) 2.94; (c) 0.337; (e) 0.032.

10.10 For $c_i \geqslant 0$ right hand side of (10.33) must be $\geqslant 0$ and using substitution provided

$$\frac{1 + \tanh^2(H\alpha/2)}{\tanh(H\alpha/2)} \geqslant \frac{1 + H^2\alpha^2/4}{H\alpha/2}$$

10.12 With $\lambda_m = 4000 \text{ km}$, $H = 10 \text{ km}$, $\partial\bar{u}/\partial z = 2 \text{ m s}^{-1} \text{ km}^{-1}$, $(kc_i)^{-1} \simeq 2.10^5 \text{ s}$.

10.15 62.2 m s^{-1}.

10.16 2×10^{-6}: 4 m s^{-1}.

10.18 $353, 475, 857, 1462, 2216 \text{ K}$; 200 days.

10.20 $u = \dfrac{-u_0[\exp(D/H) - \exp(z/H)] \sin(y/L)}{[\exp(D/H) - 1]}$;

$w = \dfrac{Lu_0[1 - \cos(y/L)] \exp(z/H)}{\alpha\beta H[\exp(D/H) - 1]}$.

11.3 1.9×10^{-3}.

11.6 18%.

12.1 (a) 0.4%; (b) 1%; (c) 3%.

12.2 Use method of q 4.11; $\sim 1.8 \text{ K}$.

12.5 The following theorem concerning differentiation under the integral sign will be found useful:

$$\text{if} \quad F(\beta, \alpha) = \int_{\alpha}^{\beta} f(x, a) \, dx$$

$$\frac{dF}{da} = f(\beta, \alpha) \frac{d\beta}{da} - f(\alpha, a) \frac{d\alpha}{da} + \int_{\alpha}^{\beta} \frac{\partial f(x, a)}{\partial a}$$

12.6 (a) 5.8%; (b) 1.7%; (c) 0.4%.

12.8 0.25 K.

12.10 420 km.

12.11 Scattering coeff. $\sim 4.10^{-10} \, \text{cm}^{-1}$; absorption coeff. $10^{-7} \, \text{cm}^{-1}$.

12.12 (a) 9.5 km; (b) 19.9 km.

12.13 Condition for ray to emerge is that horizontal ray distant r from centre of planet shall have radius of curvature $> r$. From information in problem 12.12 this condition is found to be

$$-\frac{dn}{dr} < \frac{n}{r}$$

For Venus atmosphere, assuming temperature of ~ 410 K, highest pressure 300 kPa.

12.14 4.5×10^{-5} to 4.5×10^{-2} kPa.

13.1 -26 K.

13.3 (a) 12 000 B.P., $+7\%$, -7%
       120 000 B.P., -4%, $+4\%$
   (b) 12 000 B.P., -1.7%, $+7.8\%$
       120 000 B.P., $+1.4\%$, -6.1%

13.4 About 4 years.

13.5 0.14% assuming albedo of 0.3.

13.6 $T_c^4 = (1 - A)T_0^4$;
     $T_1^4 = 2(1 - A)T_0^4$.

13.7 0.25 W m^{-2}; 0.15 K.

13.8 0.7 K.

13.10 Last part for the three cases
       (i) 11, 16, 23;
       (ii) 9, 16, 20;
       (iii) error 0.0001 at Y_4, then $n = 13, 18, 22$.

Index

absorption
  bands, 41, 240–51
  by carbon dioxide, 173, 247–51
  by single lines, 41
  coefficient, 10, 40–1
  cross-section, 66
  spectra, 10, 42
  spectrum of oxygen, 240
  spectrum of ozone, 40, 141, 250–1
  by water vapour, 42, 172, 243–7
absorptivity, of clouds, 83, 86
acoustic cut-off frequency, 108
acoustic waves, 108, 169
adiabatic
  lapse rate, 4–5
  super-adiabatic condition, 6
  wet, 19–23
air, thermodynamic data for, 224
airglow, 70, 78
albedo, 2, 51
  of clouds, 83
  for single scattering, 83
angular momentum
  of atmosphere, 162
  transport of, 153–6
  zonal, 29
anticyclone, 92, 154
argon, 58, 226
atmosphere
  composition of, 1, 226
  middle, 57–79, 156
  model atmospheres, 227–36
  plane parallel, 10
  predictability of, 210 ff
  unit of pressure in, 1
  upper, 57–79, 236
available potential energy, 24–8, 138, 153
  eddy, 29, 138, 153

available potential energy *continued*
  generation of, 28
  zonal, 29, 138, 153

back-scatter ultraviolet spectrometer, 200
balloons, 190–1
band
  electronic, 9, 240–1
  models, 49–50
  pure rotation, 9, 243
  vibration–rotation, 9, 41, 243–51
baroclinic, 122
  instability, 140, 147–50
  model, 166
  waves, 138
baroclinicity vector, 122
barotropic, 145
  atmosphere, 165
  instability, 145–7
  model, 165
beta-plane approximation, 113, 115, 147
Bjerknes–Jeffreys theorem, 102
black-body radiation, 9, 11
  from sun, 10, 239
black-body function (*see also* Planck function), 73
blocking patterns, 218
Boltzmann factor, 71
Bouguet's law, 10
boundary layer, 127, 129
  in numerical models, 176–7
Boussinesq approximation, 115, 140, 185
Bowen ratio, 136
broadening
  collision, 41, 54
  Doppler, 41, 53, 79
Brunt–Vaisala frequency, 8, 106, 109, 119

carbon dioxide
    absorption by, 42, 173, 247–50
    and climatic change, 216, 219
    collisional relaxation in, 74, 79
    concentration of, 226
    emission, 48–9, 73, 194–7
    and non LTE, 73, 78
centrifugal acceleration, 90, 97
chlorine in stratosphere, 68, 79
cirrus clouds, 80
Clausius Clapeyron equation, 21, 85
climate
    of past, 212
    variations, 212
    modelling, 216
climatic change, 1, 212–3
clouds, 80–7
    cirrus, 80
    cumulus, 80
    emissivity of, 83, 174
    formation of, 80
    interaction with radiation, 50, 82–3, 214, 221
    lenticular or wave, 80, 110
    in numerical models, 174–5
    particles, 80–2
    satellite pictures of, 33, 34, 36, 101, 193
    stratus, 80, 83, 86
    on Venus, 3, 18, 87
coalescence, in clouds, 81–2
collision
    broadening, 41, 54
    excitation of molecules by, 71
    relaxation by, 71, 73, 79
composition
    of atmosphere, 226
    of thermosphere, 58
computational instability, 168
condensation
    and formation of precipitation, 80–2
    in models, 171
    level, 20
constant flux layer, 131
continuity, equation of, 95, 102, 107
continuum absorption, 50, 55
convection, 10, 13, 14, 34, 36, 101
    in numerical models, 169
    in a rotating fluid, 138
    sloping, 138, 140, 150
convective adjustment, 169
cooling to space approximation, 48, 73, 74
Coriolis parameter, variation with latitude, 91, 112
Coriolis term, 90, 92, 93, 97
COSPAR reference atmosphere, 62, 63, 66, 236

critical level, 59, 64, 158
cumulus clouds, 34, 80
Curtis–Godson approximation, 45, 54
Curtis matrix, 75
curve of growth, 43–4
cyclone, 92, 154
    mid-latitude, 150
    tropical, 93
cyclostrophic motion, 93, 103

Dalton's law, 58
data
    assimilation in models, 178
    for weather forecasts, 206
derivative
    partial or local, 88
    total, 88
dew point, 22
diffusive separation, 59
dishpan experiment, 139
dissociation rate, 67
Dobson–Brewer circulation, 70, 79, 159
Doppler broadening, 41, 53, 79
drag coefficient, 76, 176

Eady wave, 149–50, 161–2
eddy
    available potential energy, 24–9, 138, 152–3
    diffusion coefficients, 175
    kinetic energy, 30, 138, 152–3
    stresses, 129
    transfer coefficients, 136, 175
    viscosity, 129
Einstein coefficients, 71
Ekman layer, 132
Ekman pumping, 132–3
Ekman spiral, 130
El Nino phenomenon, 215
Elsasser band model, 49, 195
emission of radiation, 10, 11
    by carbon dioxide, 48–9, 73, 194–7
    by clouds, 82–3
    by planet, 2
    stimulated and spontaneous, 71
    by water vapour, 173
emissivity
    of water vapour, 173
    of clouds, 83, 174
    of surface, 176, 193
energy
    available potential, 24–9, 138, 152–3
    balance models, 223
    budget, 48, 152
    eddy kinetic, 30, 138, 152–3

energy *continued*
  internal, 23–4
  kinetic, 24, 28
  total potential, 23–8
  transport, 151
  zonal kinetic, 152–3
entropy, 19–22
  of dry air, 19
equation
  of continuity, 95, 96, 102, 107
  of motion, 91
  of state, 3, 19, 24
  thermal wind, 94, 100, 148
equivalent width, 43–4, 54
  of collision broadened line, 54
Ertel potential vorticity, 123–4
exosphere, 59

feedback processes in atmosphere, 213, 221
Fick's law of diffusion, 81
finite differences, 167
flux
  of hydrogen atoms, 64
  radiative, 12, 47
  solar, 2, 40
fog, 31, 33, 194
forecast errors, 211
forecasting models, 177–9
frame of reference, 88
  inertial, 89
  rotating, 89
friction
  effect on geostrophic approx., 92
  in the boundary layer, 132
  Rayleigh, 141
  velocity, 131
front, 99, 101
frost point, 22

general circulation, 28, 138–64
geopotential, 94
  height, 94, 99, 227
geostrophic
  approximation, 92
  wind, 92, 97, 102, 114–15, 129
  quasi-geostrophic approximation, 115, 168
  zonal velocity, 143
Global Atmospheric Research Program, 210–11
  first global experiment, 189
Goody random model, 49
gradient wind, 93, 98
gravity waves, 8, 106–10, 119–20, 157–8, 162–3, 177, 184

greenhouse effect, 3, 15
  runaway, 16
  on Venus, 3, 16
grey atmosphere approximation, 10, 12

Hadley circulation, 143–4, 151
Hartley band of zone, 40
heating rate, radiative, 48, 75, 171
helium, 59, 76, 226
homopause, 58
homosphere, 58
hydrogen, 226
  in the upper atmosphere, 59–65
hydrostatic equation, 3, 30, 107
  for moist air, 20
hydrostatic equilibrium, 24

ice
  in glaciers, 212
  particles in clouds, 81, 86
inertial
  flow, 97
  frame of reference, 89
  instability, 143–4
  oscillations, 118
instability, 143
  baroclinic, 140, 147–50
  barotropic, 145–7
  computational, 168
  inertial, 143–4
  latent, 32
  potential, 31
integral equation of transfer, 45–7
integrated absorptance, 43
interhemispheric circulation
  in mesosphere, 158, 163
internal energy, 23, 24
intertropical convergence zone, 147
intrinsic frequency of gravity waves, 109
inversion, of temperature profile, 4
isentropic surfaces, 25–7
isobaric
  surface, 27
  system, 96
isobars, 22, 100

jet stream, 100
  polar front, 140
  subtropical, 140
Jupiter
  great red spot, 216–17
  internal energy, source in, 2, 7
  negative viscosity situation in, 156

Jupiter *continued*
   radiative parameters for, 2
   structure of atmosphere, 1

Kelvin waves, 121
kinetic energy, 24, 28
   eddy, 30, 138, 152–3
   zonal, 30, 152–3
Kirchhoff's law, 11
Kolmogoroff law, 136

Ladenberg and Reiche function, 54
Lagrangian flow, 183
Lambert's law, 10
laminar flow, 126
lapse rate, 13
   adiabatic, 4, 5
latent heat
   of condensation, 20, 171, 225
   of fusion, 225
   of sublimation of ice, 86
latent instability, 32
lee waves, 33, 34
lenticular clouds, 80
line strength, 41, 44, 241

Malkmus model, 55
Mars
   cloud formation in, 155
   greenhouse effect for, 16
   negative viscosity situation in, 156
   radiative parameters for, 2
   radiative time constant in, 17
   structure of atmosphere, 1
mass action, law of, 67
mesopause, 57, 157
mesoscale model, 179
mesosphere, 40, 48, 57, 157–8
meteor trails, 191
methane, 63, 226
microwave measurements, 202
middle atmosphere, 57
   general circulation in, 156
   Programme, 159
Milankovitch, climate theory, 219–20
millibar, unit of pressure, 1
mixing
   in lower atmosphere, 58
   length, 131
   ratio of water vapour, 20, 227
molecular diffusion, 58, 65
momentum, absolute, 185
Monin–Oboukhov length, 136

Nimbus
   satellite, 196, 198, 199
nitric oxide in stratosphere, 69
nitrogen, 1, 59, 226
noise equivalent power (NEP), 208
noise equivalent temperature (NET), 208
nuclei
   for condensation, 80
   hygroscopic, 86
numerical modelling, 165–88
   baroclinic models, 166–7
   barotropic models, 165
   of climate, 216
   data requirements for, 206
   primitive equations for, 168–9
   spectral, 180
   of transfer across the surface, 176–7
   two dimensional, 180

occultation of radio signals by Venus, 208
optical depth, 11, 13, 83
optical path, 11
orography
   cloud formation due to, 80
   in numerical models, 169
oxygen
   absorption by, 57, 65, 240
   atomic, 59, 63
   isotopic composition of, 212
   microwave band of, 194, 197
   origin of atmospheric, 77
   photochemistry of, 66
ozone, 40, 52, 57, 66–7, 77
   absorption by, 40–1, 171, 240–1, 250–1
   distribution of, 52, 68
   heating rate by, 52
   remote sounding of, 200
   spectral band information for, 250–1

partial derivative, 88
perturbation method, 105, 107
photochemical processes, 66–70
   in numerical models, 180
photodissociation, 66
physical constants (table of), 224
Planck function (*see also* black-body function),
     47, 174, 192, 236–7
planetary waves, 111–12, 159
polar front jet, 140
polarization of sunlight, 52
potential energy, 23
   available, 24–9, 138, 152–3
   total, 23–8
potential instability, 31

potential temperature, 19, 22, 25, 27, 35, 107
potential vorticity, 114
predictability
   atmospheric, 210 ff, 222
   first and second kinds, 215–16
   short-term, 210
pressure modulator radiometer, 198–9
primitive equation models, 168–9

quasi-geostrophic approximation, 115, 168
quasi-horizontal motion, 91

radar measurements, 37, 191
radiance, 195
radiation
   black-body, 9, 11
   budget, 50, 51
   infrared, 2, 9, 192–3
   solar, 2, 10, 40–1, 50–1, 171–4, 192, 199, 220,
      240–1
   terrestrial, 51, 171–4
   ultraviolet, 57, 199
radiative
   balance of earth, 2, 221
   equilibrium model, 9–18, 156
   time constant, 15, 17
   transfer, 10, 46–9, 70, 171
   transfer in clouds, 83
   transfer in thermosphere, 65
radiosonde, 190
   ascent, 30
random band model, 49, 55
Rayleigh friction, 141
Rayleigh scattering, 39–40, 52, 208
Rayleigh's criterion, 147
reaction rates, 66, 78
   for three-body collisions, 66, 78
refractive index
   in spherical atmosphere, 208
   of dry air, 225
relative humidity, 20
relaxation, collisional, 71, 73, 79
remote platforms, 203
remote sounding from satellites, 192
   of atmospheric temperature, 123, 192–8
   of atmospheric composition, 199–201
   of temperature of Venus, 103
   from atmospheric limb, 201, 208
Reynold's number, 126
Reynold's stresses, 127–9
Richardson number, 134, 137
   flux Richardson number, 137
rocket sondes, 190
rocket trails, 58, 110, 111

Rossby number, 93, 97, 160
Rossby waves, 111, 113, 115, 120
roughness length, 131

Sändstrom's theorem, 5
satellites, 76, 192, 204
   geostationary, 192, 204
   images from, 33, 34, 36, 101, 193
   polar orbiting, 192, 204
saturation mixing ratio, 20, 22
saturation vapour pressure, 20, 225–6
scale height, 3, 59
   in the thermosphere, 59
scattering
   in clouds, 83, 84
   coefficient, 83
   isotropic, 84
   of solar radiation, 39
Schwarzchild's equation, 11, 46
sea breeze front, 187
selective chopper radiometer, 198
sigma co-ordinates, 169, 170–83
sloping convection, 138, 140, 150
solar constant, 2, 238
solar wind, 65
sound waves, 105
source function, 73
spectral
   band information, 241–51
   energy density of turbulence, 133–4
   lines, 41
   numerical model, 180
spin-down time, 133, 136
stability (*see also* instability)
   for vertical motion of dry air, 5, 7
   of moist air, 21, 101
standard pressure, 19, 224
static stability parameter, 166
steering level for Eady waves, 149
Stefan–Boltzmann constant, 2, 224
Stefan–Boltzmann law, 11
Stokes' law, 86
stratopause, 57
stratosphere, 15, 40, 48, 66, 124, 157–9
   energy transport in, 152–3
   photochemical processes in, 66–70
   temperature structure of, 60–1, 157–9, 199,
      227–35
   water vapour in, 79, 159, 227
stratospheric warming, 116
stratus clouds, 80, 83, 86
streak photographs, 139
stream function, 113
strong approximation, 43, 45, 54

sub grid scale processes, 175
sub-tropical jet, 140
supercooled droplets, 80
supersaturation, 86
symmetric circulation, 140–3
synoptic scale, 91, 97

temperature
  discontinuity at surface, 13, 17
  distribution of atmospheric, 60–1, 227–35
  entropy plot, 22
  equilibrium, 2, 7
  inversion, 4
  of past century, 212
  potential, 19, 22, 25, 35, 166
  sea-surface, 215
  virtual, 30
tephigram, 31, 32, 35
terrestrial radiation, 51, 171–4
thermal wind equation, 94, 100, 148
thermodynamic
  breakdown of LTE, 70–6
  cycle, 5, 6
  diagram, 22
  engine, 5
  equation, 166
  equilibrium, 11
  local thermodynamic equilibrium (LTE), 11, 70
thermodynamics, 19–38
  first law of, 4, 19
thermosphere, 57, 59, 65
thickness, 95
Tiros satellite, 36, 192, 203, 205
tornado, 95
total derivative, 88
trade winds, 143, 145
transmission
  of an atmospheric path, 44, 75
  average, 44–5
  of clouds, 83, 87
  fractional, 11
  of slab, 48–9
transmissivity, of clouds, 83, 87
tropical cyclone, 93
tropopause, 14, 57, 159
troposphere, 14
tropospheric forcing, 153
turbidity, Angström factor, 52
turbopause, 58, 65
turbulence, 126–37
  spectrum of atmospheric, 133–5
  three-dimensional, 134
  two-dimensional, 134

Upper Atmosphere Research Satellite, 202

Venus
  cloud albedo of, 87
  density measurements for, 208
  cloud cover of, 3, 18
  greenhouse effect on, 3, 16
  lower atmosphere of, 6
  radiative parameters for, 2
  remote sounding of temperature, 103
  structure of atmosphere of, 1, 16, 103, 209
virtual temperature, 30
viscosity
  eddy, 129
  kinematic, 126
  molecular, 126, 129
  negative, 156
von Karman's constant, 131, 136
vorticity, 112, 114, 120, 132
  absolute, 112, 145,
  equation, 102, 113, 148, 165, 167
  potential, 114, 123

water vapour
  absorption by, 42, 172, 243–7
  continuum, 50, 55
  dimer, 50, 55
  distribution, 227
  emissivity of, 172–3
  and greenhouse effect, 15–16
  in numerical models, 172–3
  properties of, 225–6
  spectrum, 42, 243–7
  in upper atmosphere, 63, 227
  and vertical motion of moist air, 20
wave clouds, 80
wave equation, 106
wavenumber, 40, 116
waves
  acoustic, 108, 169
  atmospheric, 105–25
  baroclinic, 138
  Eady, 149–50
  external, 108
  gravity, 106–10, 119–20, 157–8, 162–3, 177, 184
  internal, 108
  Kelvin, 121
  lee, 109, 110
  planetary, 111–12, 159
  Rossby, 111, 113, 115, 120
weak approximation, 43, 45, 54
weighting function, 195, 198, 207
wet adiabatic, 22, 23

wind
  in boundary layer, 127–31
  geostrophic, 92, 97, 102, 114–15, 129
  gradient, 97
  mean zonal, 62
  thermal, 94
  trade, 143, 145
  in thermosphere, 111
window region, 50, 55, 196, 205

World Climate Research Programme, 215
World Weather Watch, 203

zenith angle, solar, 220
zonal
  angular momentum, 29
  available potential energy, 24–8, 138, 153
  kinetic energy, 30, 152–3